1982

List of titles

Already published

Cell Differentiation	J.M. Ashworth
Biochemical Genetics	R.A. Woods
Functions of Biological Membranes	M. Davies
Cellular Development	D. Garrod
Brain Biochemistry	H.S. Bachelard
Immunochemistry	M.W. Steward
The Selectivity of Drugs	A. Albert
Biomechanics	R. McN. Alexander
Molecular Virology	T.H. Pennington, D.A. Ritchie
Hormone Action	A. Malkinson
Cellular Recognition	M.F. Greaves
Cytogenetics of Man and other Animals	A. McDermott
RNA Biosynthesis	R.H. Burdon
Protein Biosynthesis	A.E. Smith

In preparation

The Cell Cycle	S. Shall
Biological Energy Transduction	C. Jones
Control of Enzyme Activity	P. Cohen
Metabolic Control	R. Denton, C.I. Pogson
Polysaccharides	D.A. Rees
Microbial Metabolism	H. Dalton
Microbial Taxonomy	D. Jones
Molecular Evolution	W. Fitch
A Biochemical Approach to Nutrition	R.A. Freedland
Metal Ions in Biology	P.M. Harrison, R. Hoare
Nitrogen Metabolism in Plants and Microorganisms	A.P. Sims
Cellular Immunology	D. Katz
Muscle	R.M. Simmons
Xenobiotics	D.V. Parke
Plant Cytogenetics	D.M. Moore
Human Genetics	J.H. Edwards
Population Genetics	L.M. Cook
Membrane Biogenesis	J. Haslam
Biochemical Systematics	J.H. Harborne
Biochemical Pharmacology	B.A. Callingham
Insect Biochemistry	H.H. Rees

OUTLINE STUDIES IN BIOLOGY

Editor's Foreword

The student of biological science in his final years as an undergraduate and his first years as a graduate is expected to gain some familiarity with current research at the frontiers of his discipline. New research work is published in a perplexing diversity of publications and is inevitably concerned with the minutiae of the subject. The sheer number of research journals and papers also causes confusion and difficulties of assimilation. Review articles usually presuppose a background knowledge of the field and are inevitably rather restricted in scope. There is thus a need for short but authoritative introductions to those areas of modern biological research which are either dealt with in standard introductory textbooks or are not dealt with in sufficient detail to enable the student to go on from them to read scholarly reviews with profit. This series of books is designed to satisfy this need. The authors have been asked to produce a brief outline of their subject assuming that their readers will have read and remembered much of a standard introductory textbook of biology. This outline then sets out to provide by building on this basis, the conceptual framework within which modern research work is progressing and aims to give the reader an indication of the problems, both conceptual and practical, which must be overcome if progress is to be maintained. We hope that students will go on to read the more detailed reviews and articles to which reference is made with a greater insight and understanding of how they fit into the overall scheme of modern research effort and may thus be helped to choose where to make their own contribution to this effort. These books are guidebooks, not textbooks. Modern research pays scant regard for the academic divisions into which biological teaching and introductory textbooks must, to a certain extent, be divided. We have thus concentrated in this series on providing guides to those areas which fall between, or which involve, several different academic disciplines. It is here that the gap between the textbook and the research paper is widest and where the need for guidance is greatest. In so doing we hope to have extended or supplemented but not supplanted main texts, and to have given students assistance in seeing how modern biological research is progressing, while at the same time providing a foundation for self help in the achievement of successful examination results.

J. M. Ashworth, Professor of Biology, University of Essex.

Cellular Recognition

M. F. Greaves

Imperial Cancer Research Fund,
Tumour Immunology Unit,
University College, London

LONDON
CHAPMAN AND HALL

A Halsted Press Book

JOHN WILEY & SONS, INC., NEW YORK

*First published in 1975
by Chapman and Hall Ltd
11 New Fetter Lane, London EC4P 4EE
© 1975 M.F. Greaves
Typeset by Preface Ltd, Salisbury, Wilts. and
printed in Great Britain by William Clowes & Sons Ltd.,
London, Colchester and Beccles*

ISBN 0 412 13110 2

Distributed in the U.S.A.
by Halsted Press, a Division
of John Wiley & Sons, Inc. New York

Library of Congress Cataloging in Publication Data

Greaves, Melvyn F
    Cellular recognition.

    (Outline studies in biology)
    1. Cell interaction. 2. Cellular recognition.
I. Title
QH604.2.G7   1975   575.8'76   75-38515
ISBN 0-470-15211-7

Contents

1 Biological communication and recognition

1.1 Signal coding

The genetic code embodies structural and functional potentiality and in differentiated cells the DNA can be compared to a punch tape that is programmed to delineate the cell's specialized activity. In order for this specific potential to be expressed in tune with the dynamic demands of the cell's environment, the biochemical pathways from gene to performance must be in contact with, and receptive to, *extracellular signals*.

This relationship exists at several levels; the integrity of an individual animal, plant or micro-organism within its total environment and the effective function of component parts all depend upon multiple regulatory controls or signals which govern and integrate the behaviour of cells, tissues, organs and individuals. Thus, while the performance potential of any given part is predetermined, the expression of this intrinsic programme is integrated into, and largely subservient to, the needs of the whole organism and occasionally the species. We may casually accept this as a fairly obvious truism today, and perhaps fail to appreciate Claud Bernard's unique conceptual and experimental insights into this problem over 100 years ago [1].

The analogy is frequently made between cells and people, both being members of heterogeneous and complex yet integrated societies. The cancer cell can then be portrayed as the wayward rebel who is unresponsive to the 'normal' conventions of society. This altruistic principle is indeed relevant to different levels of biological organization and although in practice possibly too impersonal and impractical for man, it is not apparently so for other 'social' creatures, such as bees, ants, and termites. It is, however, interesting and not altogether unexpected to find that social behaviour of these insects, like that of cells, may be a result of dictatorship rather than true altruism.[2] .

Biological and machine-based transactions both involve regulated activity, which in cybernetic terms popularized by Norbert Wiener [3] are dependent upon information transfer and feedback control. Cause/effect and supply/demand are continually cross-checked and performance thereby evaluated and geared to meet the challenge. In order to understand how this is achieved, we must concern ourselves with both the structure and language of the intercellular communication systems.

The relevant structures are systems within systems (the Russian doll principle) and can be arranged into domains of descending size that are concerned with the receipt of information (i.e. stimuli), and its transmission and translation (i.e. into response).

The messages we receive from the outside world can be in the form or modality of sound, smell, visible and invisible (ultraviolet, infrared) light, heat, pressure; however, *all* are translated, via the externally orientated sensory receptors of the body, into the *common language* of the nervous system — the nerve impulse. The language used is 'common' in the sense that it embodies no specific instructional information content. The informational significance of nerve cell communication from the sensory organs to the brain lies in the selective activity of sensory receptors themselves and the precision and patterning of the physical connections and pathways that exist. This is no explanation of how nerve impulses can be interpreted by the brain as one of a thousand different shapes or smells, or how such specific sensory brain centres and sensory organs become connected; however, it serves to emphasize that communication is via a language which by its *rate* of transmission — rather than specific content — relates *presence* and *level* of a particular stimulus and by its contacts and circuitry communicates form or quality. The reality of this arrangement is illustrated by the capacity of light, electricity and pressure to elicit a common sensation of 'light' and of the ability of judiciously applied electric pulses or catecholamines to elicit complex motor activities. So effective is direct stimulation of the 'pleasure areas' of the brain that rats are prepared to drive themselves to neural ecstasy and eventual death!

The brain is the communication headquarters and oversees virtually all vital processes in higher organisms. It communicates indirectly with tissues of the body via the pituitary gland — the neural-hormonal coupling centre — or more directly via nerve fibres. The nerve impulse itself serves primarily to regulate the release from the nerve endings of pre-packaged *chemical messages* — neurotransmitters — or 'local hormones' whose specificity of action lies in the cellular relationship served by the nerve and the possession by the 'target' cell of appropriate 'discriminatory' binding sites analogous to the body's sensory organs. Numerous important interactions exist between other differentiated cells of the body of which those mediated by hormones are the best

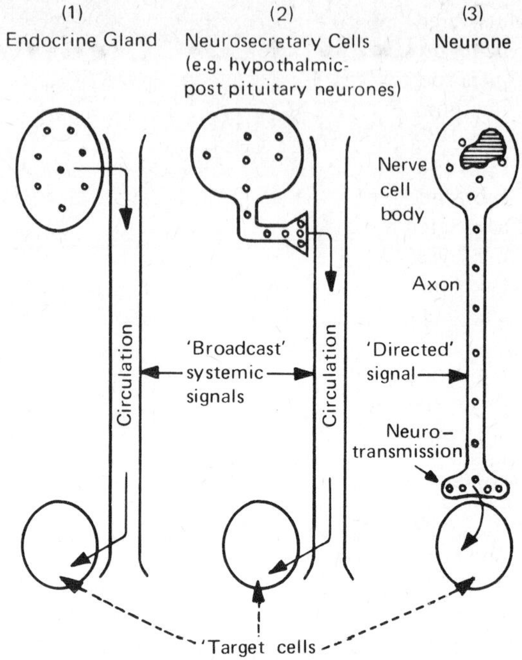

Fig. 1.1 Systemic and local 'hormones'

known and most important example. Hormones and neurotransmitters have essentially similar regulatory functions as exemplified (see Fig. 1.1) by the dual (local or systemic) role of some catecholamines (— epinephrine) and the existence of neuro-secretory cells [4].

The languages used for *inter*cellular communication are essentially all chemical and it is undoubtedly the great diversification and sophistication of this type of signal that characterizes 'internal' biological control. Diffusible chemical signals may have a zone of influence as small as a hundred angstroms (e.g. the neuro-muscular junction) or throughout the whole organism (blood-borne messages such as hormones) or extracorporeally, over a few kilometres (given a favourable wind!)

Chemical languages vary in their vocabulary and correspondingly in their specificity — 'depending on the privacy of the messages delivered and the intricacy of the transaction being proposed' [5]. They exist for inter and intra-species' communication as odours (or pheromones) which are usually, although not invariably, volatile [6], for intercellular relay both as soluble diffusable neurotransmitter substances and hormones, and as cell surface associated molecules. Intracellular chemical messages exist in the form of cyclic nucleotides (see Chapter 4) and the primary message exists encoded in the base sequences of nucleic acid.

All communications of importance in regulating biological activity involve multiple parameters often with sequential changes of language. The way in which these signals are integrated and interpreted however, at present eludes us. It is a common human experience that smells evoke salivation, whereas to a male moth, miniscule amounts of female odour can induce vigorous flight upwind in hot pursuit of sex. We are far from understanding the nature of odour discrimination, however, we can appreciate that in each example a sensory device for distinguishing between different chemicals has converted or translated this modality of information into the common language of nerve excitation and via various relays, back again into an internal chemical signal responsible for eliciting the overt physical response of salivary glands or wing muscles.

Consider also a person, not altogether uncommon, who introspects and communicates to others solely in the English language. Samuel Pepys' cyphered diaries are lost on him, as is a message in bush telegraph, smoke signal, morse, pictoglyphics, semaphore, hieroglyphics, shorthand or Portuguese. Input information is 'received' but not understood. In this sense all versatile information signalling systems are encoded in *arbitrary* units (phonemes to linguists), and are by no means the sole prerogative of 'spies'. Signals then have no intrinsic 'meaning' and their significance lies in the association they involve and the responses they elicit. Arbitrary encoding has obvious advantages:

(1) It enables as few as two units, by combination or patterning, to represent complex messages (i.e. the computer binary code).

(2) It assures privacy by reserving interpretability only to those intimate with the codes.

(3) It greatly increases the efficacy of communication over relatively large distances without loss of privacy (e.g. Napoleon's heliograph in Egypt and Nelson's semaphore at Trafalgar).

Regulatory signals in biological systems are elicitive rather than instructive (i.e. 'Darwinian' rather than 'Lamarkian'). Indeed, it could hardly be arranged otherwise, since the basic instructions for response are in the recipient's genes. The key to an understanding of the way cells 'talk' to each other lies not only in the physicochemistry of the signals themselves, but also in the means by which they are deciphered.

1.2 Signals and the cell surface

We now know that the receipt and translation of signals is largely a cell surface phenomenon and is dependent upon the existence of membrane associated 'cognitive' elements or *receptors*. In many cases, these have been directly identified and partially or completely purified, in other systems their existence is entirely hypothetical. Steroid hormones provide an important exception to this generalization. In this case, the specific receptors are intracellular and the efficacy of the steroid signal, therefore, depends crucially on its lipophilic (hydrophobic) nature which enables it to enter cells [7].

The cell surface membrane is a two-dimensional interface between a cell and its immediate extracellular environment and as such, provides the ideal venue and physical platform for interactions and signal reception. Besides maintaining the general physical and

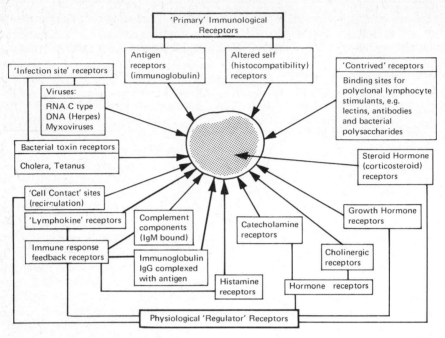

Fig. 1.2 The sensory world of lymphocytes

metabolic integrity of the cell, the surface membrane therefore serves as an elaborate sensory device capable of detecting environmental signals which induce, alter, or regulate cellular activity. Multiple receptor types can be identified on individual specialized cells, and a parallel with sensory organs of the body surface is suggested by the finding that separate receptors for different chemical signals may co-exist on the same cell and communicate intracellularly using a common chemical language. Thus, in fat cells, hepatocytes and other cell types, receptors for several different hormones exist on the cell surface and the activity of each appears to be associated with the activation of adenyl cyclase [8]. This enzyme in turn catalyses the formation of cyclic AMP – a ubiquitous small molecule which has become known as the 'second messenger'; the first messenger being the hormone or other ligand impinging upon the cell surface (see below). Functionally differentiated or specialized cells each have their own particular spectrum of surface receptors which delineate the diversity of potential environmental signals and the accessibility of the cell's performance to regulation. Fig. 1.2 illustrates a speculative but plausible view of the sensory world of lymphocytes. No two specialized cells and no two species of micro-organism or animal have the same sensory capacity – to plagiarize Von Uexkull [9] they each have their own 'Merkwelt' (perceptual world).

The way in which the cell membrane performs its function of signal recognition is not understood and represents one of the great challenges of present-day biology. The simple conceptual framework shown in Fig. 1.3 illustrates the principles likely to be involved (see also [10]).

The first component involved, is the cognitive element or receptor which functions

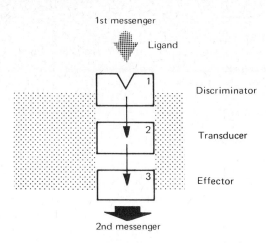

1st messenger

Ligand

Discriminator

Transducer

Effector

2nd messenger

Fig. 1.3 A model for receptor function in cell membranes (largely based on [10]).

as a *discriminator* for detecting a particular specific regulatory signal, whether this be a soluble ligand or a component of another cell surface. Two key features of receptors are therefore their asymmetric cell surface disposition and their combining site specificity, since they govern input and the range of regulatory signals which can feed into the system. From recent concepts on membrane structure it seems that outward orientation of the active binding site(s) on a receptor may be guaranteed on thermodynamic grounds, provided it is associated with hydrophilic or polar regions of the molecule. Its specificity will depend upon its more detailed chemistry as will be described below.

Receptors, however, serve for the selective receipt of signals *and* for the initiation of the cellular response. In a general sense, therefore, without prejudice as to their nature, we can consider them as bifunctional molecules, with the crucial quality of being able to communicate ligand binding to the *transducer*. The transducer is the most illusive element in the chain and in some respects the most vital, since it has the responsibility of translating the binding activity of the receptor into the

appropriate second signal. In principle, therefore, this component has a signal-response coupling function analogous to the electromagnet in many man-made communication devices. It may be part of the receptor molecule itself or a separate molecular entity (see Chapters 4 and 5).

The third vital element in information transfer across membranes is the 'effector', 'transmitter' or 'amplifier' component which is responsible for communicating altered cell surface activity into the interior of the cell. In many recognition systems (e.g. many polypeptide hormones and catecholamine neurotransmitters) this component is almost certainly the enzyme, adenyl cyclase. In other systems other plasma membrane enzymes (e.g. guanyl cyclase, Na, K dependent ATPase) or ion gating molecules may serve an equivalent effector role (see Chapter 4).

Whilst the structural *diversity* of receptors is predictable from specialization of cell function and the variety of regulatory signals, there is no *a priori* reason why transduction and effector mechanisms in cell membranes should be greatly diversified. On the contrary, it is to be expected that different cell types would employ a few common mechanisms which reflect some general properties inherent to membrane structure and function and which are to be found in phylogenetically primitive beasts. Indeed, the integrative function of biological membranes must have been a prerequisite for the diversification of cell function.

The 'second messenger' has already been identified in a great variety of cell response systems as cyclic AMP [11]. Recent evidence suggests that cyclic GMP may have an essentially similar messenger role, although one which is often antagonistic to cyclic AMP [12]. These two intracellular signals are regulated at the membrane-cytoplasm interface by inward orientated cyclase enzymes (Fig. 1.3). It is possible that a few other (but not many) analogous 'messenger' molecules exist and in principle at least these could be derived also from outside the responsive cell — their uptake being stimulated as a result of transducer activity. We should perhaps also bear in mind that receptor activity frequently results in 'turn-off' (or step down) rather than 'turn-on' (or step up) regulatory activity and, in principle at least, stimulation could result from the *reduction* of an *inhibitory* signal. It can be anticipated that these general principles may also hold true for those receptor components of the body's sense organs which are involved in the transduction of external sensory input (of various modalities) into nerve impulses. I should not be surprised if the crucial membrane events involved in neurotransmission at cholinergic synapses in the human nervous system are very similar to those occurring in the pheromone receptors of insects or even in the chemoreceptors of bacteria and protozoa. Such a situation might be anticipated on simple theoretical grounds since the receptor cells of many sense organs are embryologically part of the nervous system [13]. The analogy is further supported by recent data which suggests that cyclic nucleotides may play an important role in visual processes in the retina [14].

The biochemical identification of these three elements in membrane recognition (discriminator, transducer, and transmitter), and the resolution of the way in which they exercise their concerted function provides one of the most exciting and important pursuits of biology.

Recent basic developments in concepts of membrane structure and function (see Chapter 3) leave little doubt that an understanding of receptor function requires an integrated analysis of rapid sequential changes in the activity of mobile components of an essentially fluid cell surface membrane. This poses the classical dilemma of 'integrationist' versus 'reductionist' approaches. Obviously, receptors must be isolated and their chemistry unravelled.

However, their function only has significance and finds expression in terms of dynamic associations between receptor molecules and other membrane constituents. To paraphrase Francois Jacob [15] — membrane recognition events may be explained by the properties of the components but cannot be deduced from them.

The nature of this problem is such that interdisciplinary research is mandatory and it is gratifying to find that the cell surface is already a common meeting ground for probing biochemists, biophysicists, embryologists, immunologists, pharmacologists, and the like.

The significance of the problem can be brought home by reference to the systems to which it applies. Cellular recognition phenomena are both diverse and fundamental in living systems. They encompass sexual unions (i.e. sperm-egg in metazoan animals, pollen-stigma in flowering plants, mating types in bacteria, algae, fungi and protozoa), the development of specialized and stereotyped contact relationships in embryogenesis (particularly those involving the nervous system), the interaction of cells with neurotransmitter and hormonal signals, interaction of cells with viruses, symbionts, parasites and antigens, and finally and most formidably, the integrative functioning of the brain. Some of us believe, perhaps somewhat optimistically, that cellular recognition via the surface membrane holds the key not only to understanding the complexities of development, but also of cancer and other major human diseases.

It would be a prime example of 'Lavosier's fallacy'* to imagine that any one of these holds the key to all others, or that all cellular interactions must necessarily involve highly discriminating receptors. Nevertheless, a view of the cell surface as a transducer of intercellular signals has emerged and entrenched itself as a fundamental biological principle.

*This term was coined by Hartley apropos Lavosier's claim that all acids must contain oxygen.

2 The principle of receptor specificity

2.1 Shapes and forces

The biological efficiency of cell associated receptors depends upon ligand binding and 'transmission' qualities. Most physiological responses demand sensitivity (i.e. detection of low levels of local-or-systemic chemical regulators), selectivity, speed and reversibility. On these grounds alone, we can anticipate that the union between ligand binding sites and receptors will be non-covalent but of reasonably high affinity. This fundamental requirement can only be logically satisfied by configurational complementarity between the recogniz*er* and the recogniz*ed* — in other words by *stereospecificity*. This reasoning is based on two general considerations: Firstly, in any biological system *where multiple choice exists*, it is virtually impossible to conceive of any molecular quality, other than configurational, which would provide the opportunity for selective affinities. Secondly, we have the important precedents of enzyme-substrate and antigen-antibody reactions in solution which are recognized as being stereospecific, high affinity, non-covalent systems. We owe to Emile Fischer the concept, derived from enzyme studies, of 'Schloss und Schussel' (lock and key) which provides an everyday parallel. Analogies between enzymes, antibodies and cellular receptors have in fact dominated thought on recognition systems, although we now know that *shape* as the sole guardian for specificity is an oversimplification in all three systems.

Selective binding of ligands is determined by three qualities of both the receptor and the relevant ligand: overall molecular geometry, detailed positioning of reactive groups and configurational flexibility. The chemical bonds used by cell associated receptors will be those already familiar to physical chemists — electronic Van der Vaals forces, hydrogen bonding and coulombic or ionic interaction [16]. These bonds can not in themselves guarantee any high level of specificity. However, since the power of these attractive forces increases as the distance between interacting groups decreases, they will not be called into play unless there is (a) sufficient complementarity (of electron cloud shapes) between the reactants to overcome steric repulsive forces, and (b) suitable orientation of reactive groups in (or movable into) appropriate pairing positions.

In other words, non-specific cohesive forces are 'responsible' for the reversible receptor-ligand bond, and the sum of these subsite interactions determines the affinity or strength of binding. Steric configuration, however, is clearly the decisive factor in the specificity of receptor action. Recent concepts also emphasize the potential importance of molecular *flexibility* in receptor function; they suggest in particular that not all operationally specific recognition events involve precise combination of partners with stable complementary geometry (i.e. the straightforward Fischer 'lock and key' principle). The formation of high affinity interactions may involve a crucial

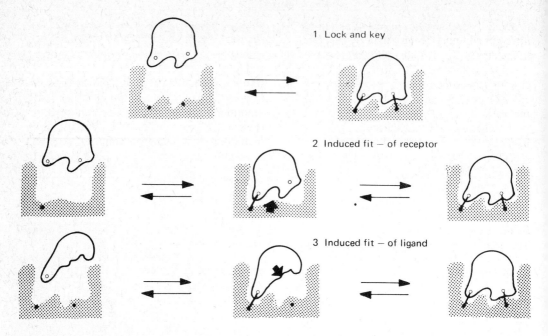

Fig. 2.1 Models for ligand-receptor interaction

element of conformational selection or induction between flexible interactants (Fig. 2.1). Binding may therefore involve changes (i) in the receptor itself (cf. the Koshland model of 'induced fit' by many enzymes [17]), (ii) in the ligand being recognized (c.f. the zipper model of double helix formation which Burgen and colleagues argue would provide a rapid on—off kinetics appropriate for cellular recognition events [18]), or both!

It is obvious that these considerations are relatively elementary and are known to be the essential features of antibody-antigen and enzyme-substrate interactions as revealed by physico-chemical studies and more recently by X-ray crystallographic analysis [169]. There is no compelling reason to suppose that membrane associated recognition events are in principle basically different, at least when they are clearly of a discriminatory nature. This is self

evident in the case of many membrane associated enzymes (some of which may in fact be involved in *inter*cellular recognition) and for cell surface membrane associated immunoglobulins which serve as receptors both on cells synthesizing these molecules (i.e. 'B' type lymphocytes) and on other cells which acquire them secondarily through the activity of receptors for immunoglobulins (e.g. mast cells). I assume the same principle to be more or less established for the cell surface receptors for neurotransmitters (acetylcholine and catecholamines) and for several hormones (particularly glucagon and insulin). Whether they are also true for other intercellular recognition phenomena is at present largely speculation and can only be considered effectively in context of the particular reaction system (see below). Of course, interactions of receptors with ligands on cell surfaces will not be as simple as recognition in solution. The kinetics of reactions will be different and micro-environmental membrane irregularities, the dynamic physical nature of membranes and the multiplicity of cell surface structures with potentially 'reactive' groups will add new dimensions to the first event in cellular recognition.

Another *a priori* consideration is the likelihood that receptors will be protein (or glycoprotein) molecules — although this need only apply to those cellular recognition systems where diversity and precise specificity are evident. The reason for this supposition is twofold: firstly, only proteins have the configurational versatility required for selective interactions, as evidenced again by enzymes and antibodies. This places the responsibility for receptor coding directly at the structural gene level and as both Monod and Jacob have argued there is a certain internal logic in this arrangement. Secondly, we know from studies on allosteric regulatory proteins in bacteria that some proteins, although not all, have the capacity to assume different configurations and to interact with other proteins or enzymes. These interactions indicate the potential means by which protein receptors could serve not only to bind ligands selectively, but also how they might initiate the 'transduction' or stimulatory process [19].

The picture that emerges is of receptors as bifunctional molecules with one (or more) active binding site(s) and additional sites responsible for 'communicating' the binding event to the translocational machinery of the cell membrane. If these are the *a priori* considerations, how well do the results of studies on cellular receptors conform to our expectations?

2.2 Historical perspective

In 1878 Langley concluded from a study on the opposing actions of atropine and pilocarpine on saliva flow in cats that — 'We may I think,

without much rashness, assume that there is some substance or substances in the nerve endings or gland cells with which both atropine and pilocarpine are capable of forming compounds. On this assumption, then, the atropine and pilocarpine compounds are formed according to some law of which their relative mass and chemical affinity for the substance are factors'. Here we have the 'germ' of receptor concept — the existence of binding substances on (or in) the 'target' cells capable of interacting with ligands (drugs in this case), according to the law of mass action.

The father figure of receptors was, undoubtedly, Paul Ehrlich who made enormous practical contributions to pharmacology, immunology and pathology, and was in fact the first to define receptors.

Ehrlich's concept of receptors derived from immunological studies on toxin anti-toxin interactions [20]. These were interpreted to imply the existence of certain haptophore groupings on toxin molecules distinct from the groupings responsible for toxicity (toxophore). Since anti-toxin could specifically interact with these haptophores, Ehrlich surmized that the sensitivity and *selectivity* of target cells might be explained by the existence of anti-haptophore groupings on these cells. The anti-toxin ('anticorps') response was then simply an excess production of these same entities into the blood. The anti-toxins were envisaged as natural *side-chains* (receptors) of the cell which normally functioned to bind, and therefore introduce, essential nutrients to the cell. The toxin was bound to the same site since it fortuitously possessed a 'haptophore grouping' corresponding (in shape?) to that of the food-stuffs. Haptophore binding thereby facilitated *selective* toxophore expression. Figs. 2.1 and 2.2 are copies of Ehrlich's diagrams from his famous Croonian lecture given to the Royal Society in London in 1910 [20].

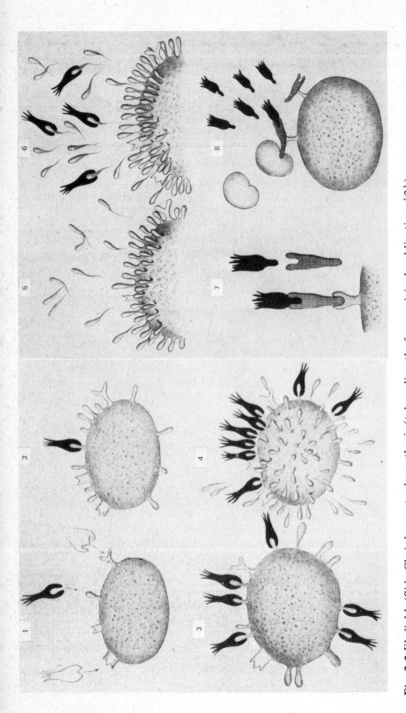

Fig. 2.2 Ehrlich's 'Side-Chain' recepter hypothesis (taken directly from original publication, [2]).

1. 'The groups [the haptophore group of the side-chain of the cell and that of the food-stuff or the toxin] must be adapted to one another, e.g., as male and female screw (Pasteur), or as lock and key (E. Fischer).'

2. '. . . the first stage in the toxic action must be regarded as being the union of the toxin by means of its haptophore group to a special side-chain of the cell protoplasm.'

3. 'The side-chain involved, so long as the union lasts, cannot exercise its normal, physiological, nutritive function . . .'

4. 'We are therefore now concerned with a defect which, according to the principles so ably worked out by . . . Weigert, is . . . [overcorrected] by regeneration.'

5. '. . . the antitoxins represent nothing more than the side-chains, reproduced in excess during regeneration and therefore pushed off from the protoplasm – thus coming to exist in a free state.'

6. [The free side-chains (circulating antitoxins) unite with the toxins and thus protect the cell.]

7. '. . . two haptophore groups must be ascribed to the 'immune-body' [haemolytic amboceptor], one having a strong affinity for a corresponding haptophore group of the red blood corpuscles, . . . and another . . . which . . . becomes united with the 'complement', . . .'

8. 'If a cell . . . has, with the assistance of an appropriate side-chain, fixed to itself a giant [protein] molecule . . . there is provided [only] one of the conditions essential for the cell nourishment. Such . . . molecules . . . are not available until . . . they have been split into smaller fragments. This will be . . . attained if . . . the 'tentacle' . . . possesses . . . a second haptophore group adapted to

The essential ingredients of Ehrlich's remarkably prophetic hypothesis were therefore:

(1) The existence, prior to toxin exposure of receptor molecules capable of interacting with the toxin.

(2) The cell surface disposition of receptors.

(3) Receptor-toxin interaction based on steric complementarity. The important analogy was made between cellular receptor-ligand interaction and reactions of enzymes and substrates which Emile Fischer had earlier attributed to a 'schloss und schlussel' (lock and key) phenomenon. Interestingly, Ehrlich also acknowledges Pasteur's notion of complementarity.

We should add (perhaps with relief!) that Ehrlich was 'off' in one respect. Although the antibody forming cell precursor can bind tetanus and cholera toxin via cell surface antibody-like (immunoglubulin) receptors, the biologically relevant binding sites for these toxins are now known to be gangliosides — the charged membrane glycosphingolipids (see Chapter 7). We also know, of course, that brain cells do not secrete antibody as Ehrlich implied. Ehrlich would, however, be consoled by the fact that his view of bacterial toxins as bifunctional molecules ('haptophore' plus binding 'toxophore') is almost certainly essentially correct.

Ehrlich's side-chain theory, as it is often referred to, not only introduced the now validated general notion of stereospecific cell surface receptors, but also provided a remarkably accurate prediction concerning the elicitation of specific antibody production. Ehrlich's picture was that of an elective rather than instructive role for antigen in inducing the production of corresponding antibodies. This is essentially the central point of Burnet's clonal selection theory of antibody synthesis proposed some half a century later [21]. Burnet introduced the important and now accepted idea of *clonal* restriction of receptor expression and antibody production. However, Ehrlich deserves much credit for predicting that antibody to an antigen is essentially an overproduction of 'receptors' by those same cells (or their descendents) that initially bound the antigen. Interestingly, Ehrlich's concept was disregarded by immunologists prior to the introduction of Burnet's theory, principally because of Landsteiner's classical work on synthetic antigens [22]. This had suggested that the immune system could respond *specifically* to an almost infinite array of antigens. This seemed at the time to demand a totally unreasonable and wasteful genetic burden for an animal to produce, prior to any challenge, complementary receptors for every conceivable antigen which it might encounter in nature or in the laboratory! Consequently, the alternative hypothesis or 'Instructive Theory' was introduced, its principle exponent being Felix Haurowitz with Linus Pauling as a notable advocate [23]. This theory proposed that antigens educated cells to make the appropriate antibody, by combining with and altering in a template fashion the shape of a preformed but 'plastic' antibody, to fit that of the antigen. The cell then somehow knew how to make many more copies of the appropriately shaped antibody. A similar notion, also incorrect, was invoked at about the same time to explain the phenomenon of inducible enzyme synthesis in bacteria.

These 'Lamarkian' views were doomed although Burnet was swimming very much against the tide when he proposed his 'Darwinian' clonal selection model in 1957. However, it seemed to him at the time, and obvious to the rest of us later, that a selective, role of antigen was much more in tune with the picture of genetic control of protein structure which was emerging in the fifties. As Burnet himself has pointed out there is a nice parallel between his view of lymphocyte selection and

Lederberg's equally classical elucidation of the *selection* of bacterial mutants *in vitro*.

The evolution of ideas on cellular receptors in the post-Ehrlich period is divisible into broadly parallel, but largely independent, pathways of development in pharmocology, endocrine physiology, immunology, reproductive biology and embryology. In all of these fields, physiological and cellular phenomenology implicated the involvement of stereospecific receptors.

It was predictable that the validity of the stereospecific receptor concept was only likely to be established in situations where soluble pure ligands were available. It is therefore not surprising that much of our current understanding of receptors is based on the study of selective action of *drugs* (see 'Selectivity of Drugs' in this series by A. Albert and [24]). Classical studies by A.R. Clark in the 1920s had an important impact on the 'receptor' theory since his quantitative studies on drug action showed that binding to receptors obeyed the law of mass action, that the interactions involved were reversible (and therefore non-covalent) and that specificity of action was manifested at very low dilutions of ligand ($\sim 10^9$ M).

The availability of *structurally related* soluble compounds was also of major significance since structure-function relationships could be established. This was not, and still is not, the case for reproductive and developmental cellular interactions, however, the pharmaceutical companies owe their lucrative business, at least in part, to the fact that it is the case for drug receptors. Extensive studies in the twenties on the differential potency of drug *stereoisomers* (e.g. of morphine, atropine) provided very compelling evidence that 'sensitive' cells possess specific receptors that had an extraordinarily precise capacity to recognize molecular *shape*. Moreover, since isomers have the same general physical and chemical properties, it seemed that *steric*

complementarity was the crucial factor in drug-receptor interaction. Subsequent work on drug analogues (e.g. of acetylcholine) have confirmed that the structural requirements for activity (e.g. Muscarinic) are in most cases very exacting. A crucial additional finding of great theoretical and practical importance was that ligands structurally related to a stimulant (or agonist) could have inhibitory (or antagonist) capacity.

The impact of these observations was enormous. Since keys to vital biological locks were available, and structural-functional relationships established, rational drug (and hormone) design became a practical goal.

Anti-psychotic drugs, psychedelic or hallucinogenic drugs, nerve gas antidotes and stilboestrol (synthetic oestrogen) all owe their potency to the receptor's 'naked rule' of stereospecificity and consequently the access of cells to manipulation. At a higher level, control of insect pests by synthetic pheromones (inducing 'sexual confusion') and of weeds by synthetic plant hormones both reflect the same underlying mechanism. The same principle also applies to antigen design in immunology although it has yet to be exploited as immunotherapy. Thus, animals, or rather lymphocytes, can be 'fooled' into thinking that they have 'seen' their own collagen molecules by injection of a synthetic polypeptide (L proline — Glycine — L-proline)n, with a triple helix conformation mimicking that of collagen [25]. A parallel 'natural' phenomenon of steric mimicry is probably responsible for the induction of at least some forms of auto-immune disease (e.g. 'cross-reactions' of streptococcal antigens with HL-A antigens on cardiac muscles in rheumatic fever [26]).

In developmental biology, the 'receptor theory' has had an inevitably rougher ride. Lillie, at the beginning of the century, studied sperm-egg interactions in sea urchins, and his discovery of fertilizin and anti-fertilizin led him to propose that gametic mating occurred by a pairing process analogous to that of antigens and antibodies. This idea was rejuvenated by Tyler in the 1940s who extended the principle of cell surface antigen-antibody like reactions to other developmental systems [27]. Reproductive interactions, particularly in microorganisms and plants, now provide us with one of the few examples of cell–cell interaction where the direct evidence for molecular complementarity is fairly convincing (see Chapter 7). In 1911, Wilson reported the apparently selective 'sorting out' association of mixtures of cells from two species of differently coloured sponges. This observation has played a major role in the development of ideas on cell interactions in embryogenesis. Similarly, the study of reaggregation of cells from dissociated embryonic tissues, pioneered by Moscona, has become a standard model for analysing the development and selectivity of histiotypic associations (see Chapter 7).

Experiments of this kind have several different interpretations, the most attractive, although not necessarily the correct one (in any or all situations; see Chapter 7) is that firm associations between identical cells (self-recognition) or mixed heterologous associations (e.g. retino-tectal) are dependent upon cell surface molecular complementarity — or mutual recognition. This idea was championed in the thirties and forties by Holtfreter, Tyler, Sperry, and Paul Weiss, who pushed the immunological analogy almost to its limits. Weiss, in particular, made a forceful argument for cell surface stereospecific receptors playing a central role in most developmental phenomena [28]. He also was perhaps the first to appreciate the wider biological implications of molecular complementarity.

Introducing the problem of specificity in growth and development [28], Weiss wrote:

'We describe as 'specific' the absorption by certain compounds of certain wave lengths of light, the relation between enzymes and their substrata; the matching between egg and sperm; the action of a hormone on its end organ; the effect of genes on characters of development; the association between a parasite and its host; the immunological response to a foreign protein; the adequate response of our nervous system to a given stimulus; the acts of recognitions and evaluation, which characterise our highest mental functions. What do these various 'specificities' have in common? Are they merely superficial parallels, or does one or the other of them perhaps contain the key to the rest so that specificity in all its manifestations of life could be resolved to a single operative principle'. And concluding the same review article:

'Rather than trying to force all biological specificity in the immunological compartment we might have to consider the latter merely as a special case of a more universal biological principle, namely, *molecular key-lock configuration as a mechanism of selectivity* (Weiss' italics), whether involving enzymes, genes, growth, differentiation, drug action, immunity, sensory response, or nervous co-ordination'. During the post-Weiss era many eminent biologists have expressed the view that molecular complementarity of the antibody-antigen type holds the key to understanding the selection and coordination observed in development. Jacques Monod has written: 'I am convinced that in the end, only the shape recognizing and stereospecific binding properties of proteins will provide the key to these phenomena' [29]. Linus Pauling, who formulated the first clear theory of stereo-specific antibody-antigen interactions, also suggested: 'that complementariness of molecular structure of some sort is responsible for biological specificity in general' [30,31]. These are 'noble' statements and few would take serious issue with them. At present, however, we must confess that the direct evidence for receptor-based specificity in development is minimal and the concept remains as an attractive but unproven hypothesis (see Chapter 7, for a fuller discussion).

19

3 Membrane structure and receptor function

Any understanding of the nature of ligand-cell surface interactions depends upon an appreciation of the basic structure of the plasma membrane itself. The cell surface membrane is much more than a semi-permeable cellular exoskeleton and it is unfortunate that the myelin sheath, so often used as an example of membrane structure, has in the past given a misleading, and somewhat corrupting, impression of the cell surface as a rather inflexible, highly structured coat of the cell. The classical models of membrane structure –the Davson-Danielli tri-lamellar structure and Robertson's 'unit membrane' are no longer accepted as generally valid. Recent conceptual and experimental advances, although by no means forsaking the direct 'morphological' approach, have placed much greater emphasis on thermodynamic considerations and more functional aspects of membrane organization.

As a result, a picture of the cell surface is emerging in which physico-chemical principles and the limitations they impose are respected and structural-functional relationships if not understood, are at least approachable.

I shall not discuss the development of ideas on membrane structure or give a comprehensive overview of the current status of the problem. Several excellent detailed reviews are readily available [32,33]. Rather, I have chosen to outline what seems at present to be the most well-founded overall concept of membrane structure – the so called 'fluid mosaic' model proposed by Singer and Nicolson [34].

3.1 The 'fluid mosaic' model

Singer and Nicolson have recently proposed a new model of membrane structure which has important implications for receptor function and membrane behaviour in general. The concept has evolved from earlier membrane models of Wallach and Singer and is applicable to membranes in general (e.g. plasma membrane, mitochondrial and chloroplast membranes). Although at present it has the status of a hypothesis it is well supported by direct and indirect experimental evidence and is in accord with basic physico-chemical and thermodynamic principles. It differs fundamentally from its classical predecessors in presenting a dynamic, flexible structure. The fluid mosaic model has two key features:

(1) Membrane proteins and glycoproteins are asymmetrically inserted or embedded in the lipid bilayer.

(2) The lipid forms the 'matrix' of the membrane and provides a viscous medium which permits translational (lateral) mobility or diffusion of the protein molecules which are therefore free to interact over relatively large distances. *The principle implication of the model is that membrane fluidity may be essential to allow interactions between protein structures as an obligatory step in their function.*

This concept of membrane structure is illustrated in Fig. 3.1 and is based principally on the proposition that membrane associated proteins, like membrane phospholipids, are

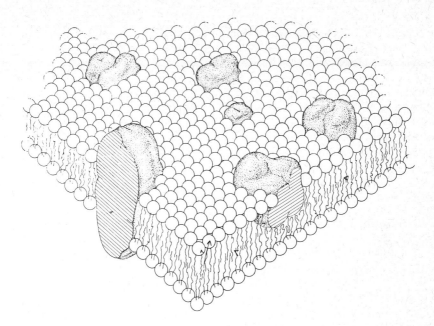

Fig. 3.1 The fluid mosaic model of membrane structure [34].

amphipathic molecules; that is to say that they have a structural asymmetry with hydrophilic polar or ionic groups and other regions with hydrophobic non-polar residues or groups. As a necessary thermodynamic consequence of this asymmetry, proteins in membranes will take up preferential orientations which reflect the lowest free energy state. This orientation would be in principle that assumed by phospholipids in films or membrane, i.e. the polar groups would be in the aqueous phase (pointed outwards) and the hydrophobic residues sequestered away from the aqueous phase in the interior of the membrane where they could interact non-covalently with hydrophobic fatty acid chains of the lipids.

As additional and critical consequences of proteins being inserted in a more or less continuous lipid matrix one would anticipate the following:

(a) that the membrane proteins would have a non-ordered or random distribution or at least there would be no long range forces intrinsic to the membrane capable of determining any large scale patterning (that is not to say that short range interactions could not result in small protein aggregates or that specialized membrane regions or domains cannot be formed, e.g. gap junctions, tight junctions, ciliary surfaces, synapses, desmosomes, acrosomal and tail regions of spermatozoa); and

(b) that proteins would be free to undergo *lateral* but not 'flip-flop' diffusion, provided they are not in some way anchored or restrained by additional forces from either outside or inside the cell. The amphipathic nature of the proteins would be expected to guarantee that no change of their perpendicular alignment, with respect to the membrane, occurred during translational movement.

The evidence for such a fluid mosaic membrane is very considerable and is presented in full elsewhere. The interested reader is recommended to read the lucid reviews by the originators themselves [33,34]. In brief, however, the principle evidence is as follows

3.2 Membrane protein asymmetry

Electron microscopic freeze etching demonstrates that a substantial amount of protein is indeed embedded in plasma membranes. Much of this protein, seen in erythrocytes and lymphocytes for example, is in the form of intramembranous particles of approximately 70–80 Å diameter, which could contain aggregates of up to a dozen or so individual molecules. Essentially similar particles have been observed in freeze-etch preparations of purified rhodopsin (the pigment protein) associated with artificial phospholipid liposomes. The number and appearance of intramembranous particles change with cell cycle and during transformation by viruses, however their relationship in nucleated cells to other proteins detectably exposed on the cell surface is at present unknown. In addition, chemical labelling studies, particularly those using 'inside-out' membranes demonstrate that some proteins may span the complete width of the membrane. To date the only well characterized cell membrane associated protein is the molecule termed glycophorin by Vincent Marchesi and colleagues [35]. This glycoprotein is a major constituent of erythrocytes. It provides binding sites for lectins and influenza virus and in addition bears blood group antigens (AB, MN). Its distribution, studied using ferritin labelled lectin, suggests that it contributes to the intramembranous particles (see above). Significantly, its carboxyl terminal end has a high content of non-polar amino acids. On the basis of these observations Marchesi and colleagues have proposed that glycophorin is orientated as shown in Fig. 3.2 – a proposition firmly in accord with the fluid-mosaic model. It is also relevant that

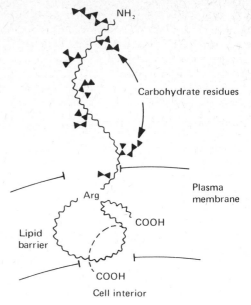

Fig. 3.2 Proposed orientation of glycophorin in erythrocyte membrane [35].

proteins isolated from membranes have appreciable α helical information. This would imply that they are likely to be globular in conformation and therefore the thickness of the membrane (75–90 Å) would demand that they are inserted relatively deeply as opposed to sitting exposed on the outside. Membrane proteins are very heterogeneous and it is highly probable that different proteins will be shown to be inserted to varying extents in the lipid bilayer. Some proteins are very easily dissociated from membranes (e.g. by increasing the ionic strength of the medium) and are considered *peripheral* in the sense that they are rather exposed and superficially associated with the membrane matrix. Most proteins in the membrane however require rather drastic treatment with detergents or organic solvents to dissociate them from membranes and we conclude therefore that these *integral* proteins are well embedded in the lipid and interact non-covalently via hydrophobic residues with the latter. Large integral proteins may span the entire membrane, other proteins may preferentially be associated with the outside or inside surface. *A priori*, one would expect receptor structures for extracellular ligands to be externally exposed, whereas plasma membrane associated enzymes with intracellular substrates (e.g. Na and K dependent ATP'ase, cyclase enzymes) might be expected to be internally orientated.

Additional evidence for asymmetric orientation of membrane glycoproteins comes from studies with lectins. The latter are sugar specific ligands (usually derived from plants of the Leguminosae family) and have been extensively used as probes for cell surface carbohydrates [36,37]. Direct electron microscopic observations on the distribution of ferritin tagged concanavalin lectin show it reacts exclusively with sugar residues (terminal α-D-mannopyranosyl or α-D-glucopryranosyl) on the *outer* surface or erythrocyte membranes. Similar findings have been reported with various lectins on a variety of mammalian cell membranes. Whilst these observations make sense in terms of receptor function, it is still not entirely clear how glycoproteins (or glycolipids) come to have their polar regions orientated into the external aqueous phase as opposed to the internal aqueous phase (i.e. the cytoplasm).

3.3 Lipid bilayer fluidity

X-ray diffraction and spin label studies (reviewed in [38]) show that the bulk of the phospholipid is probably in the form of a continuous bilayer and strongly suggest, along with results of differential calorimetry, that the lipid is in a liquid rather than crystalline form. The degree of liquidity or conversely viscosity will depend critically upon temperature (phase shifts in lipids occur at critical temperatures) and on the proportion of unsaturated fatty acid

chains of lipids. Physical studies in model (e.g. artificial membrane) systems have demonstrated that the proportions of unsaturated fatty acids influence phase transitions. Similar experiments with bacteria have demonstrated changes in the temperature dependence of transport processes with increasing concentrations of shorter unsaturated fatty acids. Two intriguing observations are relevant here. Firstly, the membrane lipids of poikilothermic organisms contain a larger fraction of unsaturated fatty acids the lower their temperature of growth — possibly to maintain sufficient fluidity for crucial membrane functions; and secondly, when some mammalian cells are stimulated by ligand-receptor interaction, increases in fluidity occur in association with enhanced phospholipid turnover and a decrease in saturated fatty acids (see Chapter 4).

3.4 Membrane protein distribution and lateral mobilization

Electron microscopic studies of fixed membranes using labelled antibodies directed against cell surface antigens indicate that most if not all cell surface antigens are randomly distributed, e.g. Rhesus antigen on red cells, H-2 and immunoglobulin molecules on lymphocytes [38]. There is still some dispute as to whether some of these antigens occur singly or in small clusters or aggregates of perhaps two to six molecules, but the results substantiate the distribution of cell surface glycoproteins assumed by the fluid mosaic model.

Two different experimental systems also provide compelling evidence that membrane antigens (presumably glycoproteins) can diffuse laterally and under some circumstances can be induced to aggregate into large patches. In an important series of experiments using heterokaryons formed by the Sendai virus induced fusion of human and mouse cells, Frye and Ediden [39] showed that immediately after fusion the mouse and human antigens are

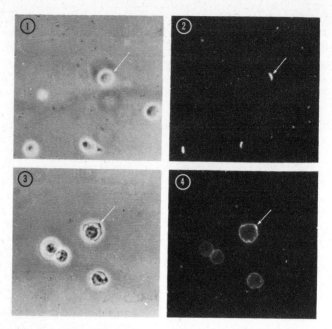

Fig. 3.3 Immunofluorescence observations on receptor redistribution (capping). 'Capped' binding sites (2), Diffuse or patchy binding sites (4), Phase contrast (1 + 3), UV light (2 + 4), [167].

largely segregated reflecting their 'native' membrane association; however, within 40 minutes they are more or less completely inter-mixed over the whole surface of the hetero-karyons. The process by which this redistri-bution of cell surface molecules takes place is presumably lateral passive diffusion, it is highly temperature-dependent but independent of protein synthesis and ATP. Other experiments on single nucleated cells demonstrate convinc-ingly that the binding of multivalent ligands to binding sites or receptor structures on the cell surface can *induce* redistribution of these mole-cules into aggregates of progressively larger size and under appropriate circumstances into a single aggregate (or 'cap') at one pole of the cell. This phenomenon was first discovered with lymphocytes and although it has been amply confirmed using many different ligand-cell inter-action systems, including fibroblasts and protozoa, it has been most thoroughly investi-gated in the context of lymphocyte cell surface immunoglobulin redistribution by anti-immuno-globulin antibodies [40]. The phenomenon is illustrated in Fig. 3.3 and its salient features are as follows:

(1) Immunoglobulins are normally diffusely distributed in the membrane as identified by monovalent antibodies (i.e. anti-immunoglo-bulins) or fixed cells treated with whole (divalent) antibody.

(2) If cells are treated with divalent antibody at 4°C the immunoglobulin becomes redistri-buted into aggregates of patches. This process occurs in the presence of metabolic poisons and is analogous to the formation of an antibody-antigen aggregate in solution.

(3) If divalent anti-immunoglobulin is added to cells at 37°C, or at 4°C and the cells subsequently warmed to 37°C, aggregates of immunoglobulin-anti-immunoglobulin become concentrated at one pole of the cell, usually the posterior or 'uropod' end with respect to the cell's directional, polarized mobility. Subsequently, the aggregated immunoglobulin is internalized by pinocytosis although it may also be directly shed from the surface.

The capping process is an active one dependent on cellular metabolism and is probably regulated by contractile cytoplasmic structures — microfilaments and microtubules which recent evidence suggests may be physically linked to the cell surface. Drugs such as cytochalasin B and colchicine which interfere with these structures can impede redistribution; however, other effects of these drugs might also be involved. The process by which a 'cap' of aggregated proteins is formed is not well understood but may relate to the backward flow of membrane on the motile lymphocyte — the aggregates become conglomerated and 'trapped' at the posterior end.

All lymphocyte surface structures so far analysed (different antigens, lectin binding sites) can be redistributed into caps; however, it is significant that few of these redistribute as readily as immunoglobulin and in many cases it is necessary to add a second multivalent layer to the reaction to achieve effective capping (e.g. anti-antibody, anti-lectin). It is presumed that this reflects a requirement for a critical degree of 'latticing' or cross-linkage of individual molecules in order for redistribution to occur. The extent to which this occurs will be determined by the surface density of the structure investigated, the valency and perhaps also the size of the interacting ligand. It should also be borne in mind that the ease of redistribution of any surface structure is also likely to depend upon its molecular associations and possible anchorage with other membrane or submembrane structures.

These observations on lateral mobility of membrane proteins have had a major influence on the plausibility of the fluid-mosaic model, and more significantly perhaps provide a lead into the key question of the relationship between membrane fluidity and receptor function. Unfortunately, to date, the cell surface molecules that have been successfully redistributed are predominantly binding sites for antibodies and lectins and although one would clearly expect all cell surface proteins to be mobile, very few studies have been reported so far on more physiologically relevant ligands (e.g. hormones, catecholamines). However, it should be noted that antigens binding selectively to the surface of those few lymphocytes bearing appropriate complementary receptors induce the same pattern of redistribution as anti-immunoglobulin. It is also important to note that cell surface binding antibodies and lectins, although perhaps not under normal circumstances functioning as physiological regulators, are in fact able to activate cells. It is therefore possible to use such systems to investigate the relationship between receptor aggregation and physiological response.

Studies with lipid soluble spin label probes and nuclear magnetic resonance (NMR) provide direct measurement of phospholipid (fatty acid chain) lateral mobility. Diffusion constants obtained were of the same order as for proteins [38]. Some lipids can probably also undergo ligand induced lateral redistribution. Thomas Revesz and I have recently [167] demonstrated aggregation and capping of lymphocyte G_{M1} ganglioside using cholera toxin which binds specifically and with very high affinity (see Chapter 7) to this glycolipid. Purified G_{M1} inserted artificially into G_{M1} deficient leukaemic cells could also be redistributed into caps.

25

4 Receptor-response coupling: a ubiquitous role for cyclic nucleotides and calcium ions?

The crucial step in receptor function is the coupling of 'recognition' to 'response'. It is only fair to say, and perhaps not surprising, that there is no cellular recognition system in which this trans-membrane signal transduction mechanism is fully understood. Nevertheless, a picture is beginning to emerge which in general substantiates the point made in the introduction — that one might anticipate the existence of a few ubiquitous mechanisms for signal transduction across membranes. In particular, it seems that cyclic nucleotides and ions (particularly calcium) may provide a common language of communication between the cell surface and the metabolic machinery of the cell.

4.1 Cyclic nucleotides and the second messenger principle

4.1.1 Cyclic AMP

The discovery by Earl Sutherland and co-workers of cyclic adenosine 3',5'-monophosphate (cyclic AMP) ranks as one of the most important achievements in biology this century and has enormous implications for receptor function [41,10,11]. cAMP was initially detected as a small stable compound in cell free liver extracts which increased the breakdown of glycogen. We now know that this nucleotide has a regulatory role in an enormous variety of cellular systems (see Table 4.1) and is ubiquitous in nature. It is found in bacteria, *Euglena*, yeast, Protozoa and all animal species investigated. Virtually all cells synthesize cyclic AMP although there are mutant bacteria and cells

('selected' *in vitro*) which are deficient in cyclic AMP mediated functions; either through a lack of cyclic AMP itself or enzymes associated with its formation or mode of action. Cyclic AMP is now known to be the chemotactic substance responsible for aggregation of vegetative amoebae of the slime mold *Dictyostelium discoideum* and may also play a crucial role in morphogenesis in the 'grex' formed from fused amoebae (see J. Ashworth's book in this series, *Cell Differentiation,* and [42,43]. These observations suggest a possible molecular basis for pattern formation and differentiation in metazoan embryogenesis.

The precise role of cyclic AMP in hormone induced metabolic responses (i.e. lipolysis, steroid production and glycogenolysis) is now well established and on a wider scale it seems likely to be of considerable importance in neurotransmission, cell division and growth control, cellular interactions and possibly also in differentiation. *The activation of cyclic AMP or its 'partner' cyclic GMP (see below) may therefore constitute one of only a few possible general mechanisms in biological systems whereby cell surface receptors exert regulatory or inductive functions*

The chemical events controlled by cyclic AMP are only well understood in two systems — the glucagon hormone regulated breakdown of glycogen by mammalian liver cells [10] and the genetic transcriptional events for inducible enzymes in *E. coli* bacteria [11]. In both systems glucose levels are of paramount

Table 4.1 Multiple regulatory roles of cyclic AMP

1. Second messenger in hormone responses
   e.g. adrenalin (e.g. liver glycogenolysis and fat lipolysis — see p. 23)
   noradrenaline (e.g. acetylcholine release in nerves)
   glucagon, ACTH, TSH, melanocyte stimulating hormone, parathyroid hormone, luteinising hormone, vasopressin, thyroxine
2. Involved in visual excitation
3. Regulates cell motility (e.g. lymphocytes, granulocytes) possibly interacting with microtubular structures
4. Regulation of cell proliferation and growth (generally suppress proliferation and facilitate expression of 'differentiated' functions — see p. 25)
5. Gene expression in bacteria (see p. 23) and eukaryotic cells [170]
6. Control of lysogeny in bacteria
7. Aggregation in the cellular slime molds (i.e. functions as a chemotactic stimulus). May also provide a morphogenic gradient regulating differentiation in these protozoa and possibly in higher animals also (in mammalian cells cAMP can pass between cells via 'gap junctions' and may be able to influence gene expression via histone kinases).

significance and the final result of cyclic AMP activity is the increased supply, via catabolic breakdown, of sugar supplies and energy.

It can be added also that the action of cyclic AMP in fat cells (mammalian adipose tissue) is in principle directed towards a similar goal since its degradative effect on stored fat (triglycerides) via the activation of lipase enzymes results in the formation of fatty acids.

It is important to bear in mind that these cyclic AMP stimulated catabolic processes are antagonized by insulin — a hormone which facilitates glycogen production and fat lipolysis and has a very different relationship to cyclic nucleotides than hormones such as glucagon (see below).

The mechanism of action of cyclic AMP in mammalian cells and *E.coli* is very different. Whereas the former involves the degradative effects of phosphorylase enzymes, the latter involves a direct influence of cyclic AMP binding protein on messenger RNA transcription (see Figs 4.1 and 4.2). Whether cyclic AMP can affect gene transcription in mammalian cells is uncertain although its capacity to increase histone nucleoprotein phosphorylation

points to a possible means by which this could be achieved.

4.1.2 Functional coupling of receptors to adenyl cyclase

Cyclic AMP is formed from magnesium-ATP, a reaction catalyzed by the enzyme adenyl cyclase (perhaps more correctly called adenylate or adenylyl cyclase). Adenyl cyclase can therefore be considered as the 'effector' molecule (cf. Fig. 1.3) in those cell surface receptor initiated events resulting in raised intracellular cyclic AMP. Significantly, this enzyme is localized in the surface membrane of cells and we presume its catalytic unit is orientated towards the cytoplasm and yet must have a special relationship with the receptor molecules themselves whose discriminator function is externally orientated (see Fig. 1.3). It has been suggested that adenyl cyclase is physically associated with hormone receptor or is even part of the receptor structure itself. Certainly some degree of co-purification of receptor and cyclase is possible and the enzyme may retain hormone responsiveness in cell free systems (e.g. small membrane fragments of cell membrane 'ghosts').

However, it is generally considered unlikely that these functionally integrated molecules are normally bound to each other or part of the same molecular structure [10,44]. As mentioned in Chapter 1 in some cell types multiple different and presumably physically separate

hormone receptors may interact with a common or shared pool of adenyl cyclase. Thus in fat cells (and their membrane 'ghosts') competitive effects on the enzyme are seen with adrenocorticotropin, epinephrine and glucagon. Secretin, and to a lesser extent, thyrotropin and luteinising hormone also stimulate adenyl cyclase in these cells. Of course, the cells are not 'universal' hormone responders and as far as is known they (or their adenyl cyclase at least) are insensitive to growth hormone, vasopressin and parathyroid hormone [10].

This situation is, however, likely to be quite common. A similar sensory 'spectrum' of adenyl cyclase has been described for cat liver and heart particles, for a mouse adrenal cortex tumour and for an astrocytoma cell line [8]. The latter example is all the more convincing for being a homogeneous, probably monoclonal, cell type. Under these circumstances it seems unlikely therefore that adenyl cyclase molecules have a physical commitment to any one receptor.

The manner in which any receptor influences adenyl cyclase is not at present known and clearly this is a key link to be discovered in the chain of receptor function. One possibility that comes immediately to mind is that adenyl cyclase might be an allosteric enzyme whose

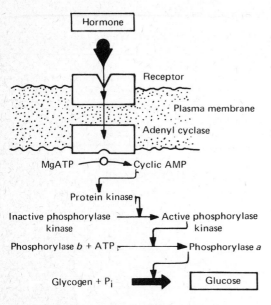

Fig. 4.1 Role of cAMP in glycogen breakdown in liver cells.

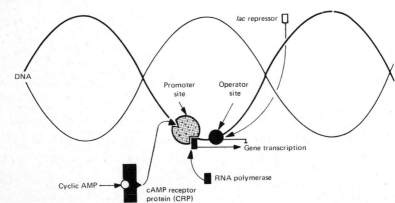

Fig.4.2 Role of cAMP in gene transcription in *E. coli*

active or more active conformation can be facilitated or stabilized by the 'regulatory' activity of the receptor. Lateral mobilization of receptors might be crucial for this interaction [45]. These interpretations imply that the receptor structure also serves as the transducer of the transmembrane signal (see Fig. 1.3). An alternative view is that the association between receptor and cyclase is indirect and enzyme responses could depend upon more 'general' or pleiotypic receptor modulated changes in membrane properties such as permeability to cations, increase in membrane fluidity and concomitant changes in phospholipids and/or effects on a particular 'transducer' molecule with affinity for adenyl cyclase.

4.1.3 Integrated control of cyclic AMP levels

Not surprisingly for a regulatory system as ubiquitous and important as cyclic AMP, the receptor – adenyl cyclase – cyclic AMP pathway is subject to numerous regulatory controls (Fig. 4.3). Thus, glucagon interaction with its receptor on liver cells is dependent upon the nucleotide guanosine triphosphate (GTP). Cyclase enzyme activity can be inhibited or in a few cases (e.g. with ACTH responsive tissues) facilitated by calcium ions. Prostaglandins [46] may serve as important allosteric

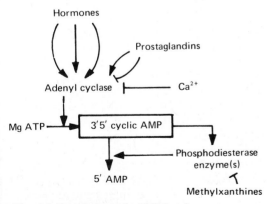

Fig. 4.3 Regulation of cAMP levels

effectors controlling adenyl cyclase activity. Although such a direct effect is not firmly established there is evidence to suggest that PGE_1 may activate catalytic activity of adenyl cyclase through interaction with a regulatory site, whereas PGA_1 may interact with an inhibitory site. Levels of cyclic AMP itself are regulated by phosphodiesterase enzyme which hydrolyses cyclic $3'-5'$ AMP to $5'$ AMP. This enzyme in turn can be inhibited by methyl zanthines (e.g. theophylline) which therefore indirectly potentiates the action of ligands operating via adenyl cyclase activation. Finally, cyclic AMP may itself induce (or increase) the formation of phosphodiesterase enzyme thus providing an 'internal' means of negative feed-back control.

4.1.4 Criteria for second messenger function

Four criteria have been proposed which should be fulfilled to establish that a hormone acts through cyclic AMP. These should be useful when cyclic nucleotide involvement is suspected in other ligand regulated cellular responses. They are as follows:

(1) The hormone should increase the concentration of cyclic AMP in the target cells or tissue.

(2) The hormone should increase the activity of adenyl cyclase in tissue extracts.

(3) Theophylline or other phosphodiesterase inhibitors (see Fig. 4.4) should mimic or potentiate the action of the hormone.

(4) Cyclic AMP and/or its derivatives should be able to mimic the action of the hormone.

These conditions are met in most of the mammalian hormonal systems investigated. The fourth criterion, involving a receptor 'by pass' response, provides the most compelling argument for a primary causal role for cyclic nucleotides in hormone action. Cyclic AMP itself is generally rather poor at producing hormone-like effects on appropriate target cells, however $N^6O^{2'}$ – dibutyrylcyclic AMP which

29

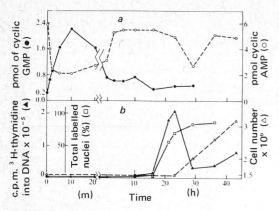

Fig. 4.4 Serum (mitogen) induced changes in cyclic nucleotide levels in mouse 3T3 fibroblasts [48]

is possibly more stable and membrane permeable than cyclic AMP itself, is very effective at inducing responses which mimic those evoked by hormones.

4.1.5 Cyclic GMP

Not all hormones appear to induce their principle effects through elevated cyclic AMP levels. In fact the response to some (e.g. insulin) may be associated with a fall in cyclic AMP concentration. It now seems likely that a second cyclic nucleotide — guanosine 3′, 5′ — monophosphate (cyclic GMP) is of considerable importance for receptor response coupling in many cellular systems [12]. Cyclic GMP formation from GTP is catalysed by guanyl cyclase enzyme, which after a good deal of controversy is now thought to be present in cell surface membranes. The reaction is potentiated by manganese ten times more effectively than by magnesium. Resting levels of cyclic GMP in most tissues are 10^{-8} to 10^{-7} moles/Kg which is at the most only 1/10th of the cyclic AMP levels. This factor has proved a stumbling block in the development of simple assay systems for cyclic GMP. However, in a

few tissues cyclic GMP levels may be considerably higher (e.g. lung and cerebellum) and in these sites the protein kinase 'target' sites for the nucleotide are identifiable.

What is the basic functional relationship between cyclic AMP and cyclic GMP? In principle, a dualism might exist with both nucleotides having positive direct effects on metabolic processes. Alternatively, it could be argued that bidirectional control exists and that cyclic AMP levels, which can be modulated directly or indirectly by cyclic GMP, provide the crucial up or down, on or off control. Nelson Goldberg of the University of Minnesota introduced the former dualistic hypothesis and has proposed that cyclic AMP and cyclic GMP operate in concert, and in general their levels will vary inversely and their effects will be antagonistic [12]. In any responding system involving cyclic nucleotides then it may be the *relative* concentrations or ratio of cyclic AMP to cyclic GMP which determines the eventual response pattern.

This intriguing idea has been compared to the Oriental dualistic concept of Yin and Yang and has aroused a great deal of interest and fervent research. To date, it is probably fair to say that whilst it is not proven to be correct there is a good deal of evidence to suggest that it is likely to be so. The criteria for establishing a role for cyclic GMP in response to any particular ligand should in principle be the same as those proposed for cyclic AMP (see p. 29) and whilst it has generally proved difficult to mimic ligand action by adding cyclic GMP or its various derivatives to cells, there is little doubt that cyclic GMP plays a crucial role in many regulatory systems [47]. The evidence is perhaps most compelling in the case of acetylcholine (e.g. cholinergic myocardial contraction), insulin (e.g. fat cell lipolysis, glycogen deposition in liver cells) and lectins (e.g. concanavalin A induced lymphocyte proliferation).

In several systems (e.g. insulin action of fat cells), cyclic GMP increases are accompanied by falls in cyclic AMP, however, this is not always the case (no drop in cyclic AMP is seen in lectin stimulated lymphocytes). It seems likely however that these two cyclic nucleotides are not independent of each other and can be directly or indirectly mutually antagonistic — perhaps, for example, via effects on phosphodiesterase enzymes or by competition for binding proteins. This mutual antagonism or reciprocity of effects is observed not only in mammalian cell systems but also in bacteria, where cyclic GMP can inhibit cyclic AMP stimulation of mRNA transcription.

Evidence for dualistic control by cyclic nucleotides seems particularly convincing in the case of fibroblast growth *in vitro*, a system which may have particular relevance to contact relationships and proliferation of malignant cells. Leifert, Rudland and colleagues have shown that various soluble ligands inducing fibroblasts to enter the cell cycle, rapidly activate guanyl cyclase and increase intracellular cyclic GMP levels [48]. Proliferation can be induced by cyclic GMP derivatives (e.g. monobutyryl cyclic GMP) themselves and throughout the cell cycle, cyclic GMP and cyclic AMP levels vary reciprocally (see Fig. 4.4). Interestingly, the dramatic reciprocal fluctuation in cyclic nucleotide levels is both rapidly induced and very transient. Exogenous cyclic AMP is capable of inhibiting growth of fibroblasts (e.g. proliferation, membrane transport processes and other 'pleiotypic' responses) and this effect can be competitively suppressed by cyclic GMP.

The inference of these and other studies is clearly that cyclic nucleotides may exert an important general regulatory control over growth, division and the expression of specialized functions. Needless to say, there has been considerable speculation about the possible importance of cyclic nucleotide changes in malignancy. Several reports suggest that exogenous cyclic AMP can restore 'normal' controlled growth characteristics to 'transformed' cells *in vitro*.

In general, the results support the view that cyclic GMP increases or facilitates the initiation of proliferation, whereas cyclic AMP increases or facilitates the expression of specialized or differentiated functions — two parameters of cell behaviour often regarded as being at least partially antagonistic.

4.2 Calcium ions and cell recognition

There is increasing evidence that calcium ions also play a crucial role in responses to many cell surface binding ligands (reviewed in [49,50]). In particular, calcium ions appear to be important for stimulus-contraction coupling (e.g. catecholaminergic smooth muscles), for stimulus-secretion coupling (e.g. nerve — impulse — neurotransmitter release, allergen — mast cell degranulation) the initiation of proliferation in numerous cell types (e.g. lymphocytes, eggs, fibroblasts, muscle cells), phagocytosis and the visual process (calcium ions are released from the outer segment discs and block sodium channels in the rod membrane, [51]). Most of these responses are thought to involve either increased influx of calcium and/or release of calcium from bound intracellular sites. Significantly, several types of response can be *directly induced by calcium ions* either by direct injection (e.g. into mast cells, or giant squid axons) or by using the ionophore A23187 — a lipophilic antibiotic which transports calcium (and other divalent cations) across membranes. A23187, in the presence of an extracellular calcium source, activates lymphocyte proliferation [52], mast cell degranulation [53], K^+ efflux from parotid glands [54] and proliferation in unfertilized eggs of sea urchin, toad and hamster [55].

In most of these situations the ionophore simulates the action of a normal physiological ligand, be this allergen (for mast cells), epine-phrine (for parotid cells) or sperms (for eggs!) and we therefore conclude that calcium ions can directly elicit all of the events crucial to these regulatory responses. Since the physio-logical ligands also operate by a calcium dependent mechanism and can in several instances be shown to stimulate calcium uptake [56], it seems reasonable to conclude that calcium ions play a central role in recep-tor-response coupling. Precisely how this role is achieved is uncertain. Receptor-ligand interac-tion may directly or indirectly open calcium gates (see Chapter 5) or may induce membrane changes which release bound calcium. Studies in artificial membrane systems suggest that release of calcium bound to phospholipids could have quite substantial local effects of fluidity and permeability of the bilayer. Once available intracellularly, calcium ions may have several regulatory actions; they may interact with contractile elements and they almost certainly interact in an important way with the cyclic nucleotide pathways. Thus calcium ions in most cases inhibit adenyl cyclase and recent studies have identified a calcium binding protein in cerebral cortex which regulates activity of phosphodiesterase enzyme (see Fig. 4.3). In contrast, guanyl cyclase is a calcium dependent enzyme and there is evidence that calcium increases cyclic GMP accumulation in lung, heart tissue and in neutrophils. Cyclic nucleotides may themselves influence calcium release and influx. In many of these situations it is difficult to sort out which is the true 'second messenger' — calcium, a cyclic nucleotide, or both! At present it would appear in terms of critical membrane transduction events that calcium may play a particularly important role in α adrenergic and cholinergic type responses and is closely tied into cyclic GMP function; the exact sequence of events however remains to be unravelled.

4.3 Pleiotypic 'early' membrane events

Ligand binding to receptors may induce ionic fluxes and cyclic nucleotide changes as dis-cussed above. These changes may occur almost instantaneously, or over a period of 30 to 60 minutes. During this period however many other membrane changes can be detected (Table 4.2). There are two significant features of these changes — they are common to many diverse systems (e.g. responses to insulin, neurotransmitters, immunological stimuli) and almost all are maximal at that concentration of ligand inducing a maximal physiological response. In other words, all seems to be relevant to the chain of events initiated by ligand-receptor interaction and all are fairly ubiquitous or common mechanisms. All of the changes listed in Table 4.2 have been observed when for example 'T' type lymphocytes are activated into proliferation by the lectin Concanavalin A, and it is clearly not easy to sort out the causal sequence of membrane events and to distinguish critical obligatory changes from others which may be 'facilitatory'.

Table 4.2 'Early' membrane changes following binding of ligands to cell surface receptors

1. Receptor redistribution
   (i.e. clustering, 'capping')
2. Conformational changes
   (e.g. fluorescence probe measurements)
3. Alteration in membrane fluidity
   (e.g. increased mobility of membrane incorporated spin labels)
4. Increased uptake of ions and metabolites
   (e.g. Ca^{2+}, K^+, P_i, sugars nucleases, amino acids)
5. Increased phospholipid turnover
   (particularly of phosphotidyl inisotol)
6. Activation of membrane bound enzymes
   (e.g. adenyl cyclase, ATPases)

(See [57, 58,] for further details).

5 Receptors for neurotransmitters and hormones

The nervous and endocrine systems constitute the central relays for intercellular communication and regulation. As outlined in Chapter 1, they are not two separate systems but rather two overlapping, highly integrated networks which share a similar mode of action on the surface of 'excitable' cells. Most of our current knowledge of receptor function is based on studies of synaptic communication via neurotransmitter molecules and of polypeptide hormones (particularly insulin and glucagon) and their action of 'target' cells. These physiological ligands interact with high affinity (association constants of 10^{-6} to 10^{-12} M) with cell surface receptors in a stereospecific manner and the major cellular responses they elicit are rapid (from milliseconds to a few minutes) and reversible.

Pharmacological and physiological evidence has established the presence of highly specific receptor mechanisms in these two systems [24,45,59]. More recently, the general strategy for analysing receptors has been to dissect the components of the receptors' 'black box' (see Fig. 5.1). The goal of these studies is to learn something of the chemical structure of receptor molecules and hopefully to elucidate receptor transduction mechanisms perhaps by re-incorporating purified receptor molecules back into simple, chemically defined artificial membranes.

5.1 Synaptic transmission
Communication between elements of the nervous system (neurones) and between nerve

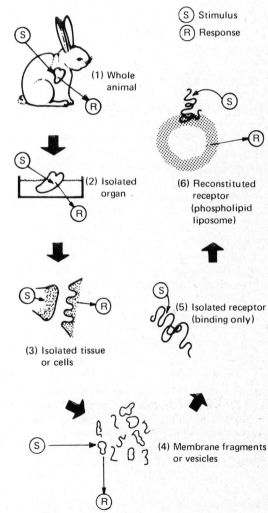

(S) Stimulus
(R) Response

(1) Whole animal

(2) Isolated organ

(3) Isolated tissue or cells

(4) Membrane fragments or vesicles

(5) Isolated receptor (binding only)

(6) Reconstituted receptor (phospholipid liposome)

Fig. 5.1 A strategy for analysis of receptors

33

terminals and end organs (e.g. muscles, glands) is primarily via transmission of chemical signals across juxtapositioned specialized cell surface regions which together constitute the synapse (see Bachelard's *Brain Biochemistry* in this series and [60]). This cell surface region is characterized by dense accumulation of membrane particles in both pre and post synaptic membranes (as revealed by electron microscopic freeze-etching studies) similar to those observed in gap junctions and tight junctions in other cell communication systems. Although a gap or synaptic cleft exists between the cell surface, some degree of direct physical linkage probably exists since synaptic junctions can withstand quite drastic mechanical and chemical (organic solvent, detergent) disruption procedures. The post-synaptic membrane is highly convoluted but it may well be a relatively rigid or paracrystalline structure — in contrast to the more fluid state of other parts of the cell surface.

Direct application of drugs and neurotransmitters have strongly suggested that receptors are concentrated in the synaptic region (e.g. motor end plate of skeletal muscles). More recently, it has been possible to confirm this directly by observing the distribution of labelled antagonist of cholinergic receptor-α bungarotoxin (a snake venom protein). The toxin localization can be determined using iodinated toxin (and autoradiography) or by using fluorescent or enzyme (peroxidase) labelled antibodies to the toxin [61]. These studies revealed that receptors are densely packed on the ridges of post-synaptic membranes where they may well constitute 50% of membrane protein. This distribution corresponds with the localization of intramembraneous particles; however, the precise relationship of receptors to particles is not yet clear. In contrast to the acetylcholine receptor, acetylcholinesterase enzyme is not concentrated in the post-synaptic region. This evidence along with results of physical membrane fractionation techniques and studies with proteolytic enzymes and cell lines *in vitro*, establishes that acetylcholinesterase is not the biologically active receptor for acetylcholine as was at one time supposed.

It is interesting to note that the restricted distribution of receptors depends upon continued functional innervation; spreading of receptors over the whole surface appears to follow as a consequence of denervation.

Our current knowledge of the function and synaptic biology of neurotransmitters is based to a very large extent on two particular systems — the cholinergic vertebrate neuromuscular junction and the electric organs of the fish *Electrophorus* and *Torpedo* [62,63]. Not surprisingly, we have less detailed information on cholinergic transmission within the central nervous system (CNS) itself. However, as discussed below, drug studies have told us a good deal about catecholamine (adrenergic) function in the CNS and it seems that essentially similar mechanisms of neurotransmission are operative. Acetylcholine and noradrenaline (norepinephrine) are the only fully substantiated neurotransmitters; however several others almost certainly exist including dopamine, serotonin, adrenaline (epinephrine), dopamine, gamma aminobutyric acid, aspartic acid, histamine and glycine.

In *cholinergic* synapses like those between post-ganglionic motor nerve fibres and skeletal striated muscle the neurotransmitter molecule is principally acetylcholine. Sympathetic nerve post-ganglionic synapses with smooth muscles and glands are, in contrast, exclusively *adrenergic* and transmission is mediated by norepinephrine. These same 'target' cells are responsive to blood borne 'hormonal' norpinephrine. The common response evoked in the same 'target' cells by hormonal and electrical stimulation led T.R. Elliot (a Cambridge student) to propose the concept of chemical neurotransmitters.

Later, this brilliantly original idea was confirmed by Otto Loewi who discovered a diffusable heart stimulant derived from the vagus nerve, and by Henry Dale (who identified this as acetylcholine). Ulf Von Evler in Sweden, later isolated noradrenaline from the sympathetic nervous system [64].

There is considerable diversity of neurotransmitters, particularly within the CNS, and also some differences in their specificity and mode of action. Fig. 5.2 illustrates a broad categorization of neurotransmitter receptor mechanisms. Although this system has been proposed primarily on the basis of drug and neurotransmitter studies, it is also pertinent to hormones. This is particularly evident with the catecholamines, adrenaline (or epinephrine) and noradrenaline (or norepinephrine). Although these molecules are only produced by cells of 'neural' origin they can function either locally

(nerve terminal release) as 'adrenergic' neurotransmitters or systematically (adrenal medulla chromaffin granule release) as adrenergic 'hormones' (see Fig. 1.1). Cholinergic and adrenergic receptor mechanisms probably arose quite early in metazoan evolution [65]. Almost all of the putative vertebrate neurotransmitter substances including acetylcholine and catecholamines are found in platyhelminth worms, arthropods and echinoderms. They are however found in widely different locations and subserve many different functions. Holothurian non-visceral muscles show both nicotinic and muscarinic cholinergic sensitivity; this probably reflects the existence of two sets of receptors rather than a single set of poorly (cf. vertebrate) discriminating receptors. The lamprey is exceptional among vertebrates in having nicotinic as opposed to muscarinic heart muscle innervation.

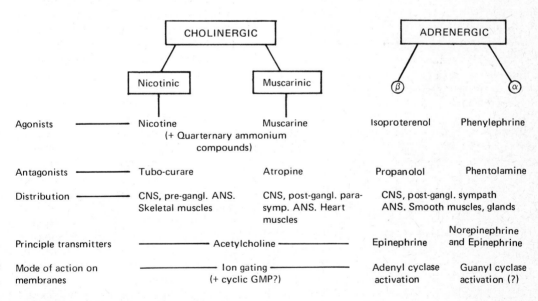

Fig. 5.2 Categories of neurotransmitters

5.2 Transduction mechanisms initiated by neurotransmitters

We assume that neurotransmitters initiate responses on excitable neurones or non-neural tissue by a purely cell surface action on receptors. The view is strongly supported by the responsiveness of membrane vesicles to acetylcholine, by the biological activity of various catecholamines covalently bound to glass beads [76] and by recent studies on reconstituted acetylcholine receptors in artificial membranes [62,63,64].

Two basic membrane transduction mechanisms appear to operate as suggested in Fig. 5.2:

(1) The rapid generation of transient membrane potential changes (in nerves and muscles) due to opening of a channel(s) for sodium and potassium. Recent studies have also suggested that calcium ions and cyclic GMP may play a role in this process, although precisely what, isn't clear! Cyclic GMP derivatives themselves induce depolarization (following after an initial hyperpolarisation) when added directly to ganglia and cyclic GMP levels can be shown to rise in sympathetic ganglia following cholinergic stimulation of pre-ganglionic nerve fibres [67].

(2) The activation of cell surface membrane adenyl cyclase which leads by an as yet unknown pathway to impulse transmission, contraction or secretion (depending on the 'target' cell). Here also a close parallel exists between neurotransmission (e.g. epinephrine) and several systemic hormones (e.g. glucagon).

A crucial question is clearly the precise relationship between neurotransmitter receptors and either the ion gating mechanism (cholinergic, α adrenergic receptors) or adenyl cyclase enzyme (β adrenergic receptors). The possible coupling mechanisms involving adenyl cyclase were discussed in the previous chapter. With respect to the ion gating, three distinct possibilities can be considered, but as yet there is no unequivocal experiment to distinguish them (Fig. 5.3). It is, however, quite possible that in the densely packed cholinergic post-synaptic membrane the acetylcholine receptor does itself constitute an ionophore or ion channel. The most suggestive evidence for this view come from experiments in which affinity purified acetylcholine receptors incorporated into lipid films or micelles (liposomes) showed cholinergic sensitivity (i.e. induced ion fluxes – [66]). The sub-unit structure of

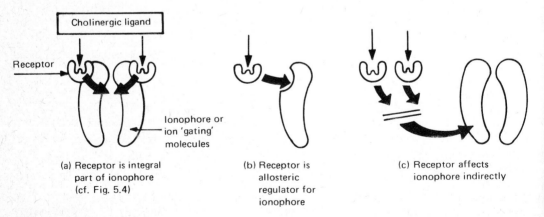

(a) Receptor is integral part of ionophore (cf. Fig. 5.4)

(b) Receptor is allosteric regulator for ionophore

(c) Receptor affects ionophore indirectly

Fig. 5.3 Possible relationships between acetylcholine receptors and the ion gating membrane mechanism

acetylcholine receptors may also accord with the provision of an ion channel (see p. 38). Any proposed mechanism should take into account the possible role of cyclic nucleotides in regulating ion gating and the involvement of calcium ion, ATP'ase (Na, K dependent) enzyme and phospholipid changes in altering membrane permeability. At present, no fully comprehensive model for either cholinergic or β adrenergic receptor function exists.

Whatever the precise molecular arrangements for transduction, it is commonly assumed that the first step in the process following ligand binding is a conformational change in the receptor itself. This view is based philosophically on extrapolation from oligomeric regulating enzyme systems in bacteria (see Chapter 2). The apparent co-operativity of responses [62,63] is often taken to support this view; unfortunately the molecular locus of the co-operative effects in responsiveness need not necessarily be at the receptor itself, and it is pertinent to note that purified acetylcholine receptors (in contrast to acetylcholine induced *responses* of cells) show no co-operation in their ligand binding activity [65]. Recent studies by Changeux and his colleagues in Paris [62] have however provided the first suggestive evidence for a conformational change in the acetylcholine receptor and it seems very plausible that such an effect plays a crucial role in signal transduction.

5.3 Neuro-receptor specificity

The specificity of cholinergic and adrenergic neurotransmitter receptors has been largely defined by comparative studies with drug *agonists* and *antagonists*. Considerable evidence shows that these ligands can directly occupy the receptor itself or in some cases may exert their effects in an indirect non-competitive manner. These same ligands have also been used extensively in attempts to map receptor active site (and subsite) regions and as affinity labels

in the purification of receptors. The existence of sterically related agonists, partial agonists (i.e. less optimal efficacy than full agonist) and antagonists which compete for the same receptor, carries the important implication that reasonably high affinity binding to the receptor is not in itself sufficient to guarantee initiation of signal transduction. One interpretation of this variable drug *efficiency* is that subtle configurational changes in receptor molecules are essential and these can only be induced by a fairly precise pattern of subsite interactions between receptor and ligand.

Studies by Karlin in particular [68] have suggested that the acetylcholine receptor has two important subsites: one involves a reducible disulphide bond and other is a negatively charged (anionic) ion pair subsite which serves as a crucial interaction site for the quarternary ammonium groups of cholinergic ligands (Fig. 5.4). Subsite groups have also been mapped tentatively for catecholamine ligand-receptor interactions [69].

Direct studies on β adrenergic receptors have been far less extensive than those on cholinergic receptors. This is partly due to the practical advantages of the latter in terms of their concentration in the vertebrate neuromuscular endplate and the electroplax of electric fish. Much of the binding observed on 'target' cells with catecholamines is certainly not to stereo-specific receptors, since at high concentrations of ligand the sum of weak non-specific inter-actions overshadows the specific binding to sparse but specific receptors [45]. Levitsky and his colleagues in Israel have however con-vincingly demonstrated stereospecific receptors for catecholamines by a method involving specific displacement of bound tritiated β adrenergic antagonist, propranolol [70]. These and other studies suggest that two general categories of binding sites exist; the biologically relevant *receptor* with association constants (i.e. affinities) of 10^{-8} M or greater and less

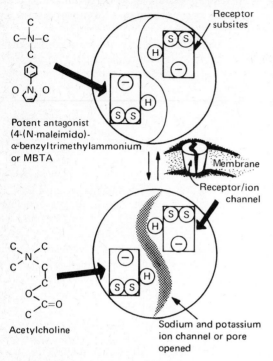

Dimeric Acetylcholine Receptor
Ionophore — viewed from above

Receptor subsites

Potent antagonist
(4-(N-maleimido)-
α-benzyltrimethylammonium
or MBTA

Membrane

Receptor/ion channel

Acetylcholine

Sodium and potassium
ion channel or pore
opened

Fig. 5.4 A model for subsite interactions in the acetylcholine receptor (after Karlin, [68]).

specific but more numerous *acceptor* sites of low affinity (10^{-4} to 10^{-7}M). Whether these latter sites can serve as true receptors by signal transduction across membranes is unknown. The existence of multiple binding sites of varying affinity and perhaps coupling efficiency provides at least a partial explanation for the observation that maximal responses can be induced both in neurotransmitter and hormone systems with apparent low receptor occupancy (1–10%).

There is some suggestive evidence that α adrenergic receptors are less specific or discriminating than β receptors [71]. The possibility that α and β receptors are in fact alternate forms of the same structures has also been discussed.

The *transient* depolarization of post-synaptic membranes is guaranteed both by the on/off nature of the nerve impulse regulated release of neurotransmitter and by mechanisms for removing or inactivating released transmitter molecules. In all vertebrate cholinergic synapses acetylcholine is rapidly hydrolysed by the enzyme acetylcholinesterase. Considerable loss of functional activity may also result from simple diffusion away from the receptor region. Although catabolic enzymes for catecholamines also exist, the 'turning-off' devices in adrenergic synapses appear to be a combination of passive diffusion plus selective *recapture* by the nerve terminals.

In contrast to vertebrate neuromuscular junctions, the body wall muscles of Ascidians have no active acetylcholinesterase. These beasts therefore contract very quickly, but relax very slowly. The motile Ascidian larvae do, as one might anticipate, show rapid, on-off, acetylcholinesterase regulated responses in their tail muscles. It is also relevant to note that some poorly reversible or covalently binding (so called 'affinity labels' [72]) ligands can induce a persistant depolarization at the vertebrate neuromuscular synapse.

Interaction of drugs with the nervous system forms the central theme of neuro-pharmacology and there is ample evidence for the enormous medical significance of the knowledge that has been gained [73]. In Fig. 5.5, I have illustrated some intriguing and important possible relationships between various drugs and catecholamine functions in the mammalian CNS [64,73]. In most cases drug action is attributable to *a close configurational similarity to natural catecholamines*. These studies provide a rationale at the receptor level for the effects of hallucinogenic drugs and have fundamental implications for both the aetiology and treatment of such neurological disorders as schizophrenia,

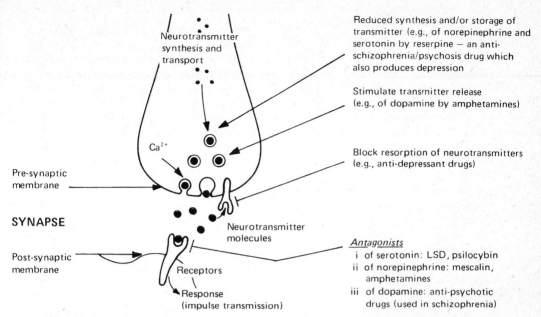

Fig. 5.5 Possible relationships between some important drugs active on the CNS and catecholamine function

Parkinsonian states, manic depression and other psychoses.

5.4 Isolation of cholinergic receptors

Very considerable progress has been made in the isolation of cholinergic receptors (reviewed in [62,63,74]. Binding sites for acetylcholine have been isolated and semi-purified from mammalian brain and striated muscle, but the richest source of these receptors is undoubtedly the electric organs of the fish *Electrophorus* and *Torpedo*.

The electroplax subunits of electric organs have a calculated receptor density of 33 000 molecules per μ^2. When this is compared with the probable density of insulin receptor on adipocytes and hepatocytes ($1-10$ per μ^2) the relative difficulties in receptor purification are self evident. In the latter case, only the equivalent of $10^{-4}\%$ of membrane protein is receptor and a 2.5 to 5.0×10^5 purification factor is required. In electric tissue the purification factor required is of the order of 100–500 fold. The cholinergic receptor density in motor end plates of skeletal muscle may approach that of the fish electroplax but since the electric organ is extremely rich in electroplax synapses this source is perhaps unique in the gross yield of receptors available.

Several distinct approaches to receptor isolation have been employed, notably by Changeux, Raftery, O'Brian, Potter, De Robertis, Rang, and their respective colleagues [74,62]. Of these, affinity chromatography (e.g. on α bungarotoxin – sepharose columns) has proved the most useful. Isolated acetylcholine receptors are extremely hydrophobic and often associated with lipid. Their stereospecificity for both agonists and antagonists is however well maintained when solubilized. Some of the properties or kinetic behaviour of solubilized receptors may be influenced by organic solvents or detergents used to obtain and maintain the receptors in a soluble phase.

In recent important experiments, reconstitution of functional membrane associated acetylcholine receptors from solubilized constituents has been achieved. De Robertis and colleagues have succeeded in incorporating cholinergic receptor rich lipoprotein into ultrathin (black) artificial lipid membranes [63]. The latter span a hole in a teflon septum separating two chambers containing solutions of ions. Using this simple apparatus it was possible to measure current voltage curves and conductance changes following the introduction of drugs, using a fine capillary tube. The introduction of receptor material itself caused a drop in membrane resistance. Significantly, a transient increase in conductance was inducible by acetylcholine and which could be blocked by (+) − tubocurarine. Similar experiments were performed with catecholamines on membranes which had incorporated adrenergic receptor lipoprotein from spleen.

Changeux and colleagues have also successfully reconstituted responsive cholinergic receptors from solubilized electroplax microsomal vesicles [63]. Recently, Michaelson, Raftery and colleagues have performed very elegant experiments in which affinity purified acetylcholine receptors were reintroduced into vesicles formed with *Torpedo* electric organ lipids. The incorporated receptor protein was directly visualized by freeze fracture and had the anticipated density of 1 molecule/2×10^5 Å^2. The majority (70%) of receptors appeared to be externally exposed and ion fluxes were induced by cholinergic ligands [65].

5.5 Hormone receptors

The attempts to identify stereospecific receptors for acetylcholine and catecholamines by binding studies and membrane fractionation have been paralleled by a similar approach to the study of hormone receptors. At the present time there is a considerable body of information on receptors for insulin [75] and glucagon [10], but much less data on other hormones [45]. Nevertheless, studies with labelled hormones already suggest that stereospecific cell surface receptors probably exist on the target cells for parathyroid hormone, follicle-stimulating hormone, luteinising hormone, prolactin, lactogenic hormone, thyrotropin releasing hormone, gonadotropin, luteinising hormone releasing factor, vaso-active intestinal polypeptide, Adrenocorticotropin (ACTH), calcitonin, oxytocin, vasopressin and growth hormone (see [45] for review).

A key breakthrough in the study of hormone receptors was the availability of relatively pure target cells (i.e. fat cell pads for insulin and liver parenchymal cells for glucagon) and radio-labelled biologically active hormones. These have enabled fairly precise correlations to be made between hormone binding and biological response. Binding characteristics of both insulin and glucagon conform to what one would expect of a stereospecific sensitive receptor system, i.e. a saturable process, specific for the hormone, and with an affinity and receptor site number consistent with physiological mechanisms. As in the case of catecholamine binding a second class of lower affinity binding ('acceptor'?) sites of questionable biological relevance has been identified. The correlation between binding and response is a crucial criterion since saturable and *apparently* stereospecific binding is demonstrable between peptide hormones and talc powder and certain glass and filter materials! The kinetic and dose-response relationships established for glucagon binding are perhaps not entirely satisfactory or at least not fully understood, since the fast linear response of adenyl cyclase contrasts with the slower non-linear (second order) kinetics of the binding process. Further convincing evidence for specific cell surface receptors is, however, derived from hormone responsiveness (of adenyl cyclase) of purified plasma membrane preparations, fat cell membrane 'ghosts', hormone binding activity of solubilized membrane proteins (see below) and biological efficacy of insolubilized hormones. In the latter studies [75] involving insulin and glucagon covalently bound to sepharose (agarose) beads of 50–300 μm diameter it is assumed that the ligand must be exerting a purely cell surface action (on specific receptors). Parallel studies have been carried out with a wide variety of cell surface binding ligands e.g. catecholamines [76], lectins [77]. A crucial point is whether the affects observed are indeed attributable solely to the covalently bound molecules, or alternatively are induced by material released from the beads. The sepharose-insulin experiments have recently been criticized on these grounds since these preparations appeared to be as effective at stimulating whole isolated fat pads (in which most cells should be inaccessible) as soluble insulin [78]. Some leaking of insulin from beads may occur and it may be very significant that this material may be polymeric and highly potent ('super-insulin' [79]). Despite these qualifications to the sepharose-ligand experiments it is still extremely likely that insulin, glucagon and other non-steroid hormones have a cell surface restricted site of action. The salient characteristics of glucagon and insulin receptors are listed in Table 5.1. Important

Table 5.1 Characteristics of hormone-receptor interactions

Characteristics	Glucagon	Insulin
1. Principle cells used for studies	Liver	Liver, fat cells
2. Receptor quantity	2.5 pmoles/mg membrane protein	0.1 pmoles/mg membrane protein
3. Binding properties		
i. Non covalent?	Yes	Yes
ii. Apparent affinity (K_m)	4.5×10^9 M	10^{-10} M
iii. 'Extra' requirements	GTP, Divalent cations?	–
iv. Do binding properties (specificity, kinetics, binding constants) correlate with with physiological activation?	Yes	Yes
4. Enzyme sensitivity of binding	Proteases	Proteases Neuraminidase + β galactosidase
5. Effect of phospholipases	Uncouples binding – adenyl cyclase response	Reveals additional functional receptors
6. Is dissociated hormone still active?	No	Yes
7. Attempts to isolate receptors		
i. Binding to membrane vesicles/fragments	Yes	Yes
ii. Solubilized receptor	Lubrol-Px: unstable complex lipoprotein	Non-ionic detergents – affinity chromatography. Insulin binding glycoprotein ($2–3 \times 10^5$ daltons)

characteristics of the glucagon receptor system, which are not evident with insulin, include the existence of a potent glucagon inactivating mechanism in rat liver, the important influence of GTP on glucagon binding and the apparently crucial role of both GTP and membrane phospholipids in the membrane coupling or tranduction mechanism (see below)

Compared to the wealth of data derived from studies of drug agonists and antagonists in the nervous system (see above) we have much less information of the precise molecular specificity of hormone receptors. The amino acid sequences of glucagon suggest that this molecule has uncharged polar, charged and hydrophobic regions which should provide subsites for interaction with the receptors active site(s) [10]. Comparisons between glucagon and secretin are of interest since these two peptides of 29 and 27 amino acids respectively are strikingly similar, yet they bind to separate receptors and only the former is able to stimulate liver cells. There are in fact only seven residue differences between these two hormones and it can be assumed that these are critically involved in the receptor's discriminator function. Four of these involve charge differences and two involve tyrosine (in glucagon) and leucine (in secretin). Removal of any of the hydrophobic residues from the carboxyterminal region results in loss of binding, whereas removal of the aminoterminal histidine residue results in loss of biological activity and the creation of a glucagon antagonist deshistidyl glucagon (with a binding affinity 1/10th that of glucagon). These studies suggest that the whole glucagon molecule is involved in its biological activity and it seems likely that binding to receptors involves hydrophobic interactions plus both hydrogen binding and electrostatic forces [10].

These subsite interactions will, as discussed in Chapter 2, be dependent upon a higher order steric complementarity between ligand and receptor and possibly on molecular-flexibility and induced conformational change. X-ray crystallography, fluorescence spectroscopy, circular dichroism and other physical studies suggest that glucagon (studied at high concentration) has a compact globular structure. However, its physical conformation may depend critically on the solvent and it is certainly possible to imagine changes occurring on binding to cell surface receptors. It has been suggested that glucagon (and ACTH) may have insufficient structural folding in solution to account for its specificity and therefore that conformational change or 'alignment' occurs in the receptor micro-environment, possibly following hydrophobic interaction of the carboxyterminal residues [10]. This idea is essentially similar to the 'zipper' model of ligand-receptor interaction discussed in Chapter 2, and it may be relevant for many small flexible peptide or polypeptide hormones. On the other hand, the studies by Hodgkin's group [80] on the crystalline structure of zinc-insulin suggest that this hormone may have all the intrinsic configuration necessary for high-affinity interaction with its receptor.

5.6 Purification of hormone receptors

Isolation of pure hormone receptors is a much more difficult task than preparing reasonable purified acetylcholine receptors. As discussed above, this difference largely reflects quantitative problems. The absolute numbers of specific receptors for both insulin and glucagon are so low (in relation to total membrane protein) that a 500 000 fold purification is required. It can be calculated that the liver homogenate of 40 rats (~ 90 gm protein) has a maximum receptor protein content of about 200 μg. These simple calculations place the problems into some kind of perspective.

Hormone receptors are integral membrane constituents and detergents or organic solvents are therefore required for their efficient extraction. Considerable success has been obtained by Cuatrecasas who extracted fat and liver cell membranes with the non-ionic detergent Triton X-100 [75]. Insulin binding macromolecules could be purified by ammonium sulphate precipitation and ion exchange chromatography (60 x purification); followed by affinity chromatography on columns of insulin derivatives (250 000 x total purification) or lectins (5000 x total purification). Affinity chromatography is the logical procedure to isolate stereospecific receptors as evidenced historically by its efficacy in purifying antibodies and enzymes using solid phase coupled complementary antigens, substrates or inhibitors (antagonists). A major problem has been the recovery of bound material in an active form. Elution in various systems can be achieved by pH or ionic changes or by affinity competition with excess soluble ligand.

Cuatrecasas succeded in obtaining 50–80% yields of insulin binding protein by elution with pH 6.0 acetate buffer containing 4.5 M urea.

Some parallel progress has been reported using affinity chromatography methods to isolate the glucagon receptor [45]. The isolated material has specific binding activity but is an unstable complex, which at present defies detailed physico-chemical analysis.

The isolated 'receptor' macromolecules have binding kinetics and specificity which are essentially similar to the intact system and are trypsin sensitive glycoproteins. They are extremely hydrophobic molecules; detergent must be present to maintain solubility and molecular weight measurements and further physico-chemical characterization is therefore difficult on both qualitative and quantitative grounds.

6 Receptors and recognition in the immune system

6.1 Immunological networks

The immunological apparatus provides the recognition system *par excellence* and embodies the ultimate sophistication of specificity, diversity and memory. The capacity of serum antibody molecules to manifest the former two properties has been known for over half a century, however it is only recently that we have begun to unravel the way in which antigen is recognized, immune responses elicited, and 'memory' established [81,82]. The great complexity of the immune system in terms of its recognition 'vocabulary', cellular hetero-geneity and regulatory controls, poses however, enormous problems for the experimenter in terms of studying physiological activation processes. Thus with any given antigen only one out of anything from 10^3 to 10^6 cells (lympho-cytes) may respond. Furthermore, the activa-tion of this cell may require collaboration with one or two other different cell types (or their soluble products); interactions which are themselves subject to complex genetic controls. The difficulties are compounded by the very considerable heterogeneity of immunologically relevant cell types (Fig. 6.1). Most, if not all, of these cells, share the property of being activated by antigen-antibody type interactions on their cell surfaces, but are programmed to respond to such stimuli in very different ways [81].

Despite the enormous wealth of cellular 'phenomenology' in immunology reflecting the multifactorial nature of immune responses, recent conceptual and technical advances have resulted in the construction of a fairly reliable outline of the organization of the immune apparatus in terms of genes, receptors, signals, cells and collaborative responses. What emerges is a view of immunological responses as a dynamic, highly integrated multicomponent surveillance system with a network of positive and negative feedback loops to regulate the behaviour of individual cellular components. Both these overall control features and the characteristics of the molecular and cellular species involved, provide a unique model for analysing biological communication and recognition.

Detailed discussions on the organization and function of the immune system can be found in two other books in this series, by M. Steward (*Immunochemistry*) and D. Katz (*Cellular Immunology*). In this chapter, I shall give only a brief overview of cell surface structures involved in immunological recognition.

6.2 Cell surface immunoglobulins as antigen receptors

Several different surface structures are involved in immunological recognition, but of these the binding site for antigen itself is the key since this is responsible for the characteristic spec-ificity of immune responses. It is now estab-lished that the cell surface molecules which 'recognize' antigen are essentially the same as antibodies (i.e. immunoglobulin [I_g] mole-cules) secreted into the blood and tissue fluid

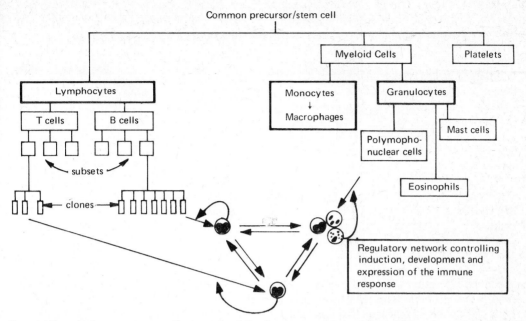

Fig. 6.1 Population structure and network of the immune system

by cells of the B lymphocyte axis (principally plasma cells). In the case of B lymphocytes themselves, these antibody-like receptors are endogenously produced and incorporated by unknown means as integral cell surface membrane proteins (see [81,83] for review). However, other immunologically active cell types can obtain specific antigen binding capacity by secondary acquisition of B cell derived antibody or can interact with antigen-antibody complexes. This is achieved by the activity of cell surface structures which recognize or bind to the Fc (carboxy terminal *F*ragment *c*rystalline of heavy chains) IgG or a similar site on IgE, and other receptors that bind complexes indirectly via serum 'complement' components. As a result of the binding of antigen or antigen-antibody complexes the cell may be activated (e.g. induction of proliferation, degranulation, phagocytosis) or some ongoing activity modulated. It is significant that B lymphocytes, bearing endogenously

synthesized antigen receptors, also, themselves have cell surface binding sites for antigen-antibody complexes and complex bound complement. These might well provide part of an important feedback control mechanism to regulate the antibody response in tune with levels of circulating antigen. Mast cells and basophils have surface binding sites for their specialized product — histamine. This also probably reflects the existence of built-in servo, or feedback system, since addition of histamine to mast cells *in vitro* suppresses antigen induced degranulation.

Antibody molecules have considerable structural diversity (see book in this series by M. Steward) reflecting both the extensive antigen combining site repertoire and the less extensive but nevertheless considerable polymorphism of the remainder of the molecule. The immunoglobulin gene pool of an individual B cell comprises at least 30 different structural genes, yet we know that from this considerable

45

store of potential diversity (over a thousand different molecules could be assembled), each *individual* plasma cell (i.e. the progeny of activated B cells) synthesizes and secretes antibody molecules which are completely homogenous in structure and specificity. Basically, only four structural genes participate in this final product – a linked pair for each type of polypeptide chain. One of each pair controls the amino-terminal sequences of the antigen combining site region, the other, the sequences of the carboxyl-terminal region concerned with 'effector' functions (i.e. binding to cells, fixing complement). This is probably a genetically unique system particularly since it involves allelic exclusion – for each pair of genes an essentially random selection of *either* maternal or paternal sets occurs. This is a form of haploid expression which is found in no other mammalian cells with the exception of the X chromosome exclusion (Barr body) in female cells. The selection of the quartet of antibody genes probably occurs early in the life of a lymphocyte and endows the cells and its clonal descendents with a uniquely specific antigen recognition capacity. Additional diversification within a clone may also occur but antibody specificity is rigidly maintained

Herein lies the whole key to the specificity of antibody responses – B lymphocytes which 'recognize' antigens only produce one receptor antibody and one (the same) antibody for bulk export out of an enormously diverse potential repertoire. This arrangement was accurately predicted by Burnet in his *clonal selection concept* [21] and explains both specific memory in the immune system (i.e. expansion of a lymphocyte clone of a particular antibody specificity) and its converse – immunological tolerance (i.e. the specific loss or inhibition of particular clones).

Immunoglobulin molecules appear to be a vertebrate invention in so far as there is no evidence for the existence of structurally related similar molecules in any invertebrates. Carbohydrate binding agglutinins are widespread in invertebrates, occurring free in haemolymph, in special organs (e.g. albumin gland of snails) or incorporated into cell membranes (in sponges) [84]. These lectin-like substances are heterogeneous in structure and usually have a very small binding site comprising less than a single hexose ring. They bind to numerous cell types including bacteria and it has been suggested, unfortunately with little or no direct evidence, that they have a protective role. While these substances are of great interest as probes for cell surface sugars (e.g. analysis of transformed cells, blood grouping) there is no obvious reason to suppose they are related in any way to immunoglobulins. The same is true for various other anti-bacterial substances found in several phyla. These may have 'protective-like' capabilities (i.e. to enhance phagocytosis) but have at the most only nominal specificity with no evidence for adaptability of response. Nevertheless, these invertebrate substances should not be neglected since similar anti-saccharide molecules could well be involved in cell-cell interactions. Recently, a very important 'clue' to the origin of immunoglobulins has been found. This will be discussed following a consideration of recognition mechanisms employed by the other major group of lymphocytes – the T cells.

6.3 Surface structures involved in T lymphocyte recognition of cellular antigens

While it is now clear that immunoglobulin molecules are responsible for stereospecific high affinity binding of antigen by B lymphocytes (and other cells passively acquiring secreted antibodies) the nature of the antigen receptor on the other major class of lymphocytes – the T cell, is shrouded in mystery and controversy (see [81] for review).

The apparent specificity and diversity of T cells in response to various antigens had earlier suggested that the T cell receptor would in all likelihood also be an immunoglobulin-like molecule. A 'minimal' hypothesis might be for example that the receptor would contain a hydrophilic, hypervariable amino-terminal end containing an antigen combining site and that at least this end of the receptor structure would be closely similar to conventional immuno-globulins. For this reason, this hypothetical structure has often been referred to as 'IgX'. This line of thought appeared to be supported by inhibition of antigen binding by T cells using anti-immunoglobulin reagents and by the direct demonstration of immunoglobulin molecules on T cells. However, over the past few years serious doubts have arisen about the repro-ducibility of some of these observations and the assumption that any immunoglobulin found on T cells was likely to have been produced by these cells is now seriously questioned. Cer-tainly under some circumstances T cells can acquire antibody molecules provided by B lymphocytes, and it may well be that in circumstances where T cells show a discrimin-atory capacity equivalent to that of free antibody molecules they are in fact using ex-free antibody! Whilst the idea of a T cell immunoglobulin has not been completely rejected, emphasis has shifted to a con-sideration of other surface structures which may represent not only T cell receptors for 'antigens', but also important sites for col-laborative interactions between T cells, B cells, and macrophages. These are cell surface glycoproteins coded for by genes that are within or linked to the major 'histo-compatibility' locus of the species, i.e. H-2 in mice, HL-A in man [85,86]. Since the nature and relevance of these molecules is discussed in detail in another book in this series (by Dr D. Katz), I shall only give a cursory or summary view of the salient points.

Although T cells collaborate in antibody responses to a great variety of antigens and can elicit very specific delayed hypersensitivity responses, the most pronounced response of this cell population is seen in interactions with other cell surface components, particularly when these are still incorporated into the surface of a living metabolically active cell. The latter might be a foreign cell (in skin grafting or 'mixed-lymphocyte' culture experiments) or alternatively, an autologous cell that had changed as a result of a disease process or through uptake and incorporation of membrane components of micro-organisms (e.g. viruses). Clearly outside of the laboratory, the latter reflects the more physiologically relevant or environmentally directed response. Reactions directed towards a single cell type, although showing signs of clonality, appear to involve a large proportion of the T cell population (say 1 in 10 compared to 1 in 10^4 to 1 in 10^6 B cells reacting to an antigen). There is increasing evidence that T cells do not show such a powerful response to all surface structures but are interested in, or programmed to respond to, cell surface structures that are coded for by genes closely linked to those defining the serological detectable major histocompatibility antigens of the species, i.e. T cells respond best to 'allogeneic' cells from members of the same species but of a different strain (or different individuals in the case of man). This genetic region is large and undoubtedly contains multiple genes coding for a variety of cell surface structures. Whilst T cells (like B lymphocytes) *may* be able to recognize all of these as antigens, it appears that some are particularly potent in activating T cells.

One group of determinants may be of particular significance. These are antigens which have two outstanding genetic characteristics:

(a) They are found predominantly on the surface membrane of macrophages and B cells (and probably on skin and sperm) but are poorly represented on T cells and other cells of the body [87].

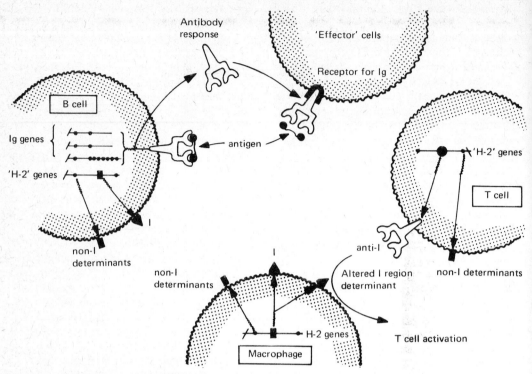

Fig. 6.2 Receptors and Recognition in the immune system

(b) They are coded for by genes which map among functionally defined, dominant genes (so called 'Ir', or Immune response, genes) which themselves determine whether or not T cell dependent immune responses (i.e. delayed hypersensitivity, T cell dependent antibody responses) can be evoked [89]. What begins to emerge is the view that the immune system embodies two distinct yet functionally integrated recognition components – the immunoglobulin antibodies produced by B cells and used also as antigen receptors – and other polymorphic cell surface structures which are concerned in the initiation of cell mediated immunity and cellular interactions (see Fig. 6.2). In this latter case it seems possible that the 'I region' in the mouse H-2 linkage group (and analogous regions in other species)

codes for structures which:

(1) provide an integral component of the T cell recognition unit – either the receptor itself or an associated molecule;

(2) are critically involved in cellular interactions between syngeneic T cells, macrophages and B cells;

(3) provide a powerful stimulant to T cells when present on a foreign (e.g. allogeneic) cell or (and perhaps most significantly) when expressed in an *altered form on an autologous cell* (e.g. a virus infected macrophage). This concept carries the intriguing implication that a chromosome region evolved and diversified so as to include and maintain sets of complementary recognition structures.

The evolutionary implications of these possible relationships are certainly considerable.

Could it be, for example, that genes for receptors involved in immunological interactions are phylogenetically related to others whose primary concern might be homo- or hetero-typic developmental cell association? Could the enormous balanced polymorphism and genetic expenditure of this region reflect some more general biological function out of which the immune system evolved as a specialized and highly successful by-product? This idea has been a favourite speculation of several eminent 'immuno'-biologists including Burnet, Bodmer, Jerne, Edelman and others [86,89] and certainly suggests what could be a very revealing experimental approach to non-immunological cell interaction systems in vertebrates (see Chapter 7). There is also a potentially fruitful analogy to be drawn between the balanced polymorphism of the histocompatibility loci in mammals and single or double loci (each with multiple alleles) in fungi and flowering plants which regulate fertilization processes through cell surface incompatibility reactions (see Section 7.1).

From the evolutionary standpoint it is intriguing to find that T cell-like specific 'graft' rejection reactions are seen in echinoderms and protochordate ascidians [90]. Allograft recognition phenomena have also been recorded in annelids and some coelenterates [84].

6.4 Origin of immunoglobulins
Recent important evidence suggests a direct evolutionary link between major histocompatibility locus coded surface structures and immunoglobulin antibodies (reviewed in [91]). Chemical studies and membrane receptor redistribution experiments (i.e. 'capping', see Chapter 3) indicate that major H-2 and HL-A molecules are associated with a small polypeptide chain (M.W. 11 000 daltons) known as β_2 microglobulin. β_2 microglobulin, like H-2 and HL-A antigens, is in fact found on the surface of virtually every cell in the body;

however, detailed amino acid sequence analysis has revealed that it has a striking homology with part of the so-called 'constant' (relatively invariant amino acids) region sequences of immunoglobulins. There is also some preliminary evidence for similar functional properties shared between β_2 microglobulin and immunoglobulin (e.g. fixation of serum complement components). It may therefore be possible that the genes coding for β_2 microglobulin, although now unlinked (in present day mammals) to immunoglobulin genes, are the descendents of genes which also gave rise to immunoglobulins. These important observations emphasize the possibility that the characteristic immunoglobulin-based diversity of the vertebrate immune system arose from a non-immunological cell surface-based system for intercellular recognition. The T cell receptor then assumes considerably expanded general interest as a possible 'intermediate' molecular form. Its formal chemical identification is awaited with great interest.

6.5 Immunological activation mechanisms
We can now therefore construct a fairly clear picture of the recognition apparatus of the immune system with its battery of cell types (B cells, macrophages, mast cells, basophils) equipped to recognize antigens via cell Ig's (endogenously or passively acquired) or in the case of T cells, armed with receptors which may recognize altered histocompatibility or differentiation antigens. We do not understand how antigen triggers any of these cell types; the membrane transduction mechanism is unknown, as is the nature of the intracellular (i.e. 2nd) messenger. Nevertheless, several vital clues are available; many of which derive from studies with ligands other than antigens which are mimetic in so far as they induce a response in individual lymphocytes which is indistinguishable from that induced by antigens [57]. These ligands are generally referred

to as polyclonal activators or polyclonal mitogens, since in contrast to antigens, they activate a large proportion of cells irrespective of clonal specificity [92,93]. A diverse group of substances including proteins (e.g. lectins and antibodies to cell surface determinants), bacterial polysaccharides, polyanions and heavy metal ions possess this capacity [81]. Significantly, anti-immunoglobulin antibodies can also be potent stimulants for mast cells and B lymphocytes. Given that a final common pathway for antigen and polyclonal ligand induced responses probably exists, the point at which the pathways converge is uncertain. It could be that these various ligands act on different membrane receptors to initiate production of a common intracellular signal, as occurs with several different hormones on, for example, fat cells (see Chapter 1 and [57]). Alternatively, the reaction pathways may be coincident from the beginning with the same primary receptor and/or 'translating' proteins being involved. Interestingly, the lectins, Concanavalin A and Wheat Germ agglutinin can mimic the effects of insulin on adipose tissue, probably through direct interaction with the insulin receptor [45]. Activation of both T cells and mast cells by lectins may also occur via interaction with binding sites on the same molecular structure, that carries recognition sites for physiological ligands (antigens) [92]. There may also be activating ligands which do by-pass the physiological receptor system(s) — for example, the lipid soluble stimulants (calcium ionophores and bacterial lipid A).

Activation of T cells, B cells, macrophages and mast cells is currently an area of intense research activity and inevitably, controversy [92,94]. Some of the generally agreed important facts to date with respect to triggering are however as follows:

(1) Initial activation is a cell surface phenomenon since it can be induced with insolubilized ligands, e.g. lectins [77], antigens [95].

(2) In all immunological triggering systems the valency of the activating ligand appears to be particularly important [57]. Many polyclonal mitogens and antigens are intrinsically polyvalent while mitogenic antibodies to lymphocyte and mast cell surface determinants must be at least divalent in order to trigger. These requirements may simply concern avidity of reaction but could also be a reflection of a necessity to cross-link and mobilize or aggregate receptors [96].

(3) As discussed in Chapter 3, calcium ions mimic the effects of physiological ligands on lymphocytes and mast cells. The pleiotypic pattern of early membrane changes has been well documented (see Table 4.2) and cyclic GMP appears along with calcium ions to play a central role in the activation process [52,56,97].

(4) Responses of all immunologically involved cells can be regulated by a wide variety of cell surface reactive ligands other than antigens (see Fig. 1.2). The precise role of the antigen specific and non-antigen specific receptors in *initiating* the response to antigen is however still uncertain, as is the interaction of different types of cells via direct contact and soluble factors. A number of speculative models have been proposed [92,94].

7 Selective cell interactions

7.1 Sexual diversification and selective gamete interactions

7.1.1 Principles

Sexual reproduction serves a basic evolutionary cause of genetic diversification, but manifests itself in an enormous diversity of forms (see [98] for review). In most single celled organisms the gametes differ little in size or structure from their vegetative progenitors; whereas in some Protista and in all higher organisms, gametes are morphologically specialized. The male-female distinction may itself be somewhat arbitrary and the gametes which form mating pairs, although different at a molecular level, may not be morphologically distinct.

Most colonial forms and many individual animals and plants are hermaphrodite, whereas all vertebrates are unisexual. Only a minority of hermaphrodite plants and animals are however self-fertile. This self-incompatibility, along with species, variant and mating-type specificity, forms the basis of an interesting and important pattern of highly selective sexual interactions. These appear, in so far as they can be (or have been) tested, to be highly discriminatory and to involve cell surface structures which serve a receptor-like role.

In fungi, no morphological differentiation into separate sexes exists and sexual inter-actions are controlled through complex systems of *incompatibility* of increasing diversification with higher forms of fungi. In lower fungi genetic control is unifactorial with a two-allele system, whereas complex bifactorial multiple allele systems are found in the high fungi (homobasidiomycetes). The diversity of incompatibility factors is insured both by the polymorphism of the few loci involved *plus* the combinational effect between the linked loci (or their products). The precise role of in-compatibility factors in 'facilitating' mating is uncertain. Successful mycelia conjugation depends upon a *lack of homologous* specifici-ties and therefore requires mating types to be different. In some systems, the factors involved are expressed on the cell surface and have been partially characterized (see below). Studies on mutants have revealed that products of incom-patibility genes serve not only for the initial contactual recognition of mycelia (i.e. code for receptors) but also regulate the morphogenetic sequence of complete sexual progression (hyphal fusion, nuclear exchange, nuclear pairing, hook cell formation, etc.). A similar and presumably related phenomenon of incompatibility occurs in many flowering plants [100,101]. Here, genetic incompatability between pollen and stylus is necessary for growth of the pollen tube and successful fertilization. In many species, incompatibility is controlled by a single locus with multiple specificities (alleles). As in fungi, the number of distinct alleles or reproductive variants can be very high (e.g. 78 in red clover, 24 in cherries). Lewis has suggested two semi-molecular models to explain the events underlying genetic control of incompatibility. In the first, a given allele codes for two distinct structures — one for the pollen and one for the style. These proteins S^p and S^{st} then have an antigen-antibody like complementarity. In the second hypothesis, the allele is complex and codes for a specific protein(s) *common* to pollen and style, plus activators for protein production. Interaction between identical protein on pollen and style results in the formation of a dimer or tetramer which functions as a growth inhibitor. The two theoretical possibilities are therefore interaction of identical chains or alternatively of steric complementary molecules — essentially the same possibilities as proposed for tissue cell interactions in embryogenesis (see below).

Significantly, these incompatibility phenomena in plants also occur in sexual reproduction of colonial coelenterates (Anthozoa) and tunicates (*Botryllus* sp.) and are remarkably similar to non-sexual 'transplantation' reactions in animals (see p. 47), revealing an important analogy first appreciated by Oka and Burnet [102]. Two characteristics are shared by these phylogenetically disparate systems. On the one hand, the remarkable genetic polymorphism resulting in the expression of multiple specificities and, secondly, the existence of cell surface interactions which reflect the apparently positive recognition of non-self. In reproductive situations the polymorphism and demand for incompatibility presumably exists to ensure heterozygosity and maximize genetic diversity by out-breeding [99]. In animal histocompatibility systems, these cell surface structures are thought to be concerned with cellular interaction in the immune response and possibly also in development either of colonies (in invertebrates) or of selective tissue associations in multicellular animals (see Chapter 6).

Even if the precise molecular biology of these cell surface entities is at present poorly understood, it is philosophically satisfying to find such strong links between sex, development and immunity. They share the common qualities of diversity and self/non-self distinction, expressed via cell surface molecules (probably glycoproteins) and may have phylogenetic (genetic-molecular) relationships as well as manifesting phenomenological similarities. Whatever its precise role, the existence of this pattern of strong genetic commitment to a balanced polymorphism of cell surface structures carries an obvious implication that some very important selective advantage is involved (see lucid discussion on this topic by Bodmer [86]).

Significantly, antigens linked to the major histocompatibility loci of mammals (e.g. H-2) are expressed on the sperm and egg cell surfaces. Genetic studies in man have suggested that the frequency of offspring that are homozygous with respect to parental haplotype specificities studied may be lower than expected by chance alone. Sperm-egg compatibility at some loci may therefore be inhibitory to successful fertilization — and accords with the situation seen in invertebrates and plants!

7.1.2 Role of the cell surface in mating type interactions of micro-organisms

One of the best studied microbial systems of 'sexual' recognition is cell fusion or agglutination between opposite mating types in the yeast *Hansenula wingei*. Marjorie Crandall and colleagues have made considerable progress in isolating cell surface reactive glycoproteins (mating 'factors') which possess demonstrable complementary affinities [103]. The mating type factors have been partially purified by chromatography (on Sephadex G-200) but may possibly be obtained in a relatively pure form by using cells of the opposite mating type as a 'cellular affinity absorbent'.

The characteristics of the Hansenula mating type factor interactions are given in Fig. 7.1. The crucial observation is that a complementary

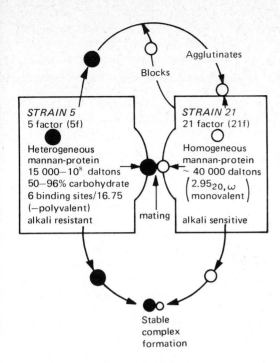

Agglutinates

Blocks

STRAIN 5	STRAIN 21
5 factor (5f)	21 factor (21f)
Heterogeneous mannan-protein 15 000–10^8 daltons 50–96% carbohydrate 6 binding sites/16.75 (–polyvalent) alkali resistant	Homogeneous mannan-protein ~ 40 000 daltons $\left(2.95_{20,\omega} \atop \text{monovalent}\right)$ alkali sensitive

mating

Stable
complex
formation

Fig. 7.1 Characteristics of mating type factors from Hansenula (fungi) [103]

interaction leading to a soluble factor complex formation is demonstrable in solution. This reaction provides a counterpart to sexual agglutination of the organisms themselves and is remarkably similar to an antigen-antibody combination. The chemistry of these factors requires further study since purified and complete gene products have not yet been isolated (factor 5 is extremely heterogeneous in molecular weight, and factor 21 is derived by proteolytic digestion); nevertheless, this system offers an opportunity to unravel the molecular basis of 'sexual' affinity between cells. It will be extremely important for example to pin down the size and chemical nature of the precise region of the 21f which is 'recognized' by 5f. Inhibition of strain 21 agglutination by 5f provides a simple and quantitative assay for this

type of analysis and parallels hapten inhibition of antigen-antibody reactions. Another key question is the nature of the membrane signals involved in complex sequence of events involved in the mating reaction — or in other words, how does stereospecific binding and agglutination initiate these membrane events crucial to subsequent genetic interaction? Are the mating type factors 'chemically' involved in this process as part of the transductional machinery (see p. 5) or are they simply 'grappling hooks' which bring together other signal initiating sites?

Similar mating type agglutination reactions occur in other yeasts. In the species *Saccharomyces cerevisae*, cellular interactions may be facilitated by the release of hormone-like factors. Mating type α, releases a polypeptide (1–2 000 daltons) which arrests the cell cycle in the complementary mating type *a*. Such a cell cycle block might facilitate mating by stabilizing the form with the optimal competence for complementary interaction and/or the subsequent mating sequence.

Mating types in Protozoa (particularly in *Paramecium*) and Algae have been extensively studied both genetically and immunologically [104,106]. The mating reaction itself is initiated by a 'recognition' reaction which is very similar to the agglutination reactions seen in yeast. Sexually competent gametes of *Paramecia* and *Chlamydomonas* for example, form stable bonded pairs via ciliary or flagella contact respectively (see Fig. 7.2a). These topographically localized contactual interactions then spread proximally to the anterior end and para-oral regions (in *Paramecia*) or the papillary regions (in *Chlamydomonas*) and the nuclear sequence of events are initiated. Some progress has been made in the isolation of mating type factors from *Chlamydomonas* by L. Wiese [105]. Mating type factors are released from gametes under certain conditions into culture supernatants. These factors or

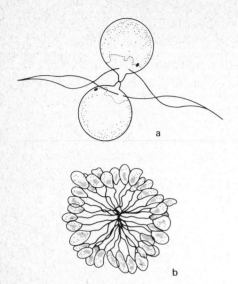

a

b

Fig. 7.2 Sexual interactions in Algae (from [105])

gamones are then capable of agglutinating gametes of the opposite sex. Addition of female mating type factors to male gametes induces *homosexual agglutination* via flagellar tip interactions (Fig. 7.2b). The gamones are therefore considered to be polyvalent isoagglutinins and are sex, species and site (i.e. flagella) specific. Since this specificity of action precisely parallels the interactions of gametes themselves, it is very probably that flagella cell surface gamones are responsible for complementary mating interaction. The isoagglutinins or gamones have not yet been prepared in a pure form. They are functionally active as glycoproteins of high molecular weight ($\sim 10^8$ daltons) which might represent either homotypic aggregates of gamone molecules or alternatively, gamone molecules incorporated in flagella membrane particles. As in the case of yeast mating factors, the precise chemical specificity of gamones is unknown as is their 'receptor' role in initiating the biochemical cascade of events involved in sexual union.

Finally, it should also be emphasized that mating interaction in different *Chlamydomonas* species (and in Algae and Protozoa in general) are extremely diverse and by no means all involve flagella interactions. In *Chlamydomonas suboogama*, for example, direct fusion of cell bodies occurs and specificity of interaction is guaranteed at least in part by chemotaxis of gametes.

The role of cilia and flagella in establishing complementary gametic contact finds a parallel in some bacterial mating processes where *sex pili* play an important role in male (+) − female (−) interactions [106]. The F-Sex-pili of *E. Coli* have been extensively studied. F-pili are coded for by genes of the sex factor (F) plasmids. The major component of the sex pili is F-pilin, a protein with one D-glucose and two phospate moieties per 12 000 dalton sub-unit and rich in amino acids with non-polar (i.e. hydrophobic) side chains. Pili probably consist of two parallel pilin rods and are about 85Å across and one micron in length. Contactual association between mating pairs occurs via pili, and pili-less mutants are 'sterile'. The precise role of pili in the mating sequence is uncertain. Since mating pairs may undergo genetic exchange with pili-only contact, it is possible as suggested by Brinton that the pili serve not only as selective contact structures but also as conveyors of DNA from male to female.

Other experiments have demonstrated an interesting selective relationship between the pili and DNA. In bacteria containing plasmids (I-like and F-like) each plasmid specifies a corresponding type of pilus. The plasmids carry different genetic markers and by selective blocking of either type of pilus (using phages or antibodies) it has been possible to demonstrate that pili preferentially transport the genetic markers coded for by the same plasmid which coded their own formation. Jacob has suggested that this selectivity is simply topographical and is determined by the proximity between the site of chromosome anchorage and that of the pilus. However, it remains possible that selective DNA transport involves an important stereospecific recognition mechanism.

Regardless of the cellular phenomenology on which preside the nuclear events in microbial mating it seems reasonable to assume that all mating interactions involve fairly discrete complementary interactions or affinity involving cell surface structures. The latter are also presumably diverse in terms of combining site specificity (although this has not been critically tested) and are glycoproteins whose existence and expression are closely linked to sexual differentiation. The structures serve as receptors in the sense that their discriminatory activity is essential for the commencement of the complex sequence of cellular and genetic events involved in mating. However, as in virtually all receptor systems, the precise role of the discriminator in the initiation of these cell surface changes which provide the signal(s) for intercellular events is unknown [104].

7.2 'Infectious' interactions on the cell surface
Infection of bacteria by viruses (bacteriophages) and of animal cells by viruses and parasitic organisms provides a fascinating example of cellular interaction. In most cases the critical first step of recognition or attachment appears to involve distinctive cell surface molecules. Burnet, who was the first to provide direct evidence for the existence of virus receptors also drew attention at that time to the possible analogy between virus-receptor and antigen-antibody reactions [107]. Whether one considers these membrane sites for infection as true receptors is a matter of semantics. They are essentially 'fortuitous' in nature and although they show discriminatory binding and are essential for subsequent fusion or penetration, they may not serve as signal transducers

in the same sense as receptors for true regulatory ligands (e.g. hormones). Cell surface binding is a key factor in determining susceptibility of cells to viral infection. Bacteria may become resistant following mutations which result in loss of the cell surface binding sites for phages [108]. Conversely, resistant cells can be rendered susceptible to infection by direct injection of virus and by incorporation of exogenous phage receptors. Multiple parameters are involved in penetration [109,110]. An initial temperature independent reversible binding to receptors (e.g. via tail fibres of phages or specialized membrane regions of parasites) is followed by a second irreversible temperature dependent step which is linked to penetration and may involve both contractile elements, enzyme activity (e.g. neurominidase in influenza viruses) and membrane fusion. There is an obvious and interesting parallel between such infectious interactions and fertilization mechanisms, particularly in the instance of virus infection which involves the transfer of genetic material. In one particular instance as discussed below, the same 'receptor' may in fact be used for both types of recognition.

In this section I shall give a brief overview on virus receptors. Unfortunately at present we have very little information on the cell surface binding sites for intracellular parasites. Some, and in particular the intracellular facultative bacteria such as pneumococcus and *Listeria*, are taken up into macrophages by pinocytosis. The species and cellular selectivity of several protozoan parasites might well be dependent upon their specificity for certain cell surface groups. Recent studies on malarial merozoite infection of erythrocytes suggest that a specialized membrane structure of the parasite is involved in binding and penetration of the red cell membrane [111] and that the binding sites on primate red cells for *Plasmodium knowlesi* merozoites are chymotrypsin and pronase sensitive (but trypsin resistant) proteins [112].

7.2.1 Bacteriophage receptors

Bacteria have more complicated surface organization than eukaryotic cells and it appears that almost every major constituent of their cell wall-membrane structure can provide binding sites for phages [108]. Knowledge of phage receptors on bacteria is at a considerably more advanced stage than on virus receptors on eukaryotic cells. This disparity reflects to a large extent the enormous advantages to be derived from strain and sero-type selectivity of different phages and the availability of phage resistant mutants.

Fig. 7.3 illustrates some of the sites of phage receptors on Gram positive and Gram negative bacteria. These studies are based partly on using purified bacterial components to inactivate phages — an approach introduced by Burnet in the 1930s and on chemical analysis of phage resistant mutants. The latter studies reveal some detailed chemistry of the receptors [108]. Thus, critical sites or linkages have been identified for several viruses, e.g. Gal-1, α-Man for E^{15}, o-acetyl group for G341, D-glucose linked to heptose for P1, N-acetyl-D-

glucosamine for Felix 0. In several instances the receptors appear to be the serologically defined determinants as might be inferred from the serotype selectivity of infection; for example, P22 phage appears to react with 12 specificity of the 0 antigen of the bacterial lipopoly-saccharide. It is interesting to note that the great majority of phages are 'lectin' like in being specific for saccharide conformations.

One of the most intriguing relationships is the function of *E. coli* F-pili as sites for both bacterial (F^+,F^-) conjugation and DNA transfer [106] and male specific phages. Single stranded RNA phages (e.g. QB, MS_2, R17) attach specifically to the sides of the F-pili whereas filamentous single stranded DNA phages (e.g. M13) attach to the tips. The direction of genetic information transfer is clearly opposite in phage infection to that in sexual conjugation. Nevertheless, this parallel provides additional support for the view that phages may have evolved from 'normal' bacterial DNA.

7.2.2 Virus receptors on animal cells

Animal viruses have selective host spectra, which in itself implies selective penetration via cell surface receptors. The capacity of plasma

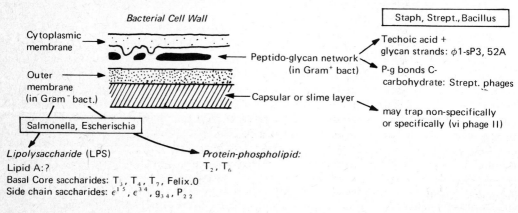

Fig. 7.3 Bacteriophage receptors on the bacteria cell surface

Table 7.1 Nature of some of the receptors for virus adsorption on animal cells

Virus	Cell	Receptor on cells	
		Chemical nature	Sensitive to
Adenovirus 2	HeLa	Lipoprotein	Subtilisin
	KB	Lipoprotein	Pronase
ECHO 7, 19	RBC	Lipoprotein	Sonication
Coxsackie B3	HeLa	Glycoprotein	Periodate
			Pronase
			Chymotrypsin
Rhinovirus 2	HeLa	Protein	Proteolytic enzymes
Poliovirus 1, ECHO 9	HeLa	Protein	Trypsin
	MK	Protein	Trypsin
	RBC	Protein	Trypsin
Reovirus 1, 2, 3	L cells	Mucoprotein	NANAase
	RBC	Mucoprotein	Periodate
Myxoviruses	FE, LU-106	Mucoprotein	NANAase
	L cells	Mucoprotein	Periodate
	Chick	Mucoprotein	

See [109] for details

membrane preparations to bind and inactivate viruses accords with this view [109,110]. Binding of viruses to cell surfaces can of course be directly observed and it seems that whereas some viruses have a preference for specialized membrane regions (e.g. cilia, microvilli coated membranes, dense reticula) others attach to less distinctive sites [110]. The chemistry of virus receptors is in its infancy, nevertheless we already have a picture of the range of molecular structures which may serve as binding sites [110,113]. In the main, these appear to be glycoproteins as judged by their susceptibility to enzymes (Table 7.1) although a fascinating experiment by A. Haywood using artificial membranes suggests that gangliosides (charged glycolipids) may provide binding sites for Sendai virus [114]. The myxo and paramyxoviruses have been extensively studied and the relevant receptors appear to be sialoglycoproteins. Their binding and penetration is dependent upon a four chain agglutinin molecule (identified as a haem — or red cell blood group, agglutinin) and the enzyme neuraminidase. The haemagglutinin appears to be specific for the acetylated sugar in α-N-acetylneuraminyl-N-acetylgalactosamine. The neuraminidase enzyme breaks the glycosidic bond between the virus and its receptor. Viruses budding from the surface of their host cells incorporate some of the latter's cell surface membrane components and it may well be that these are relevant to subsequent interactions and infectivity. Attempts to reveal incorporation of *known* lymphocyte surface antigens into budding viruses using immuno-electron microscopy have not so far been successful [115].

Many viruses have important relationships with cells of the immune system and this can result in both immunosuppression and malignancy (i.e. leukaemia or lymphomas [116]).

Several viruses appear to have a 'tropism' for lymphocyte subpopulations. For example, in man, measles virus binds selectively to 'T' lymphocytes (and suppresses T cell dependent immune responses in man and mice [117]); in contrast Epstein-Barr Virus (EBV) binds only to 'B' type lymphocytes [118]. EBV 'transforms' normal B lymphcytes into continuously dividing lymphoblastoid cell lines *in vitro* and *in vivo*, and it is probably, aside from causing infectious mononucleosis, responsible for Burkitt's lymphoma — a B cell neoplasia found associated with malarial infections in Africa [119]. EBV is a primate specific virus and it would be of considerable interest to know if its receptor was associated with any of the HL-A linked B cell surface antigenic-structures which play an important role in immune responses (see Chapter 6). Other viruses have a wide host cell range and may have a much less specific interaction with the cell surface. Pox viruses, for example, attach to many different cell types and even to any appropriately charged surfaces. [110].

Finally, recent studies using human-mouse somatic hybrids (i.e. heterokaryons formed by Sendai virus induced fusion) have provided the first attempt to map onto distinct chromosomes the genes' coding for a virus receptor [120]. Polio virus resistant lines were identified and the only consistent characteristic of these, compared with susceptible lines, was the *loss* of human chromosome *19*. It is therefore concluded that chromosome *19* contains genes that are crucial for susceptibility to polio virus, and most probably these are structural genes for the cell surface virus receptor.

7.2.3 Bacterial toxins

I include these toxins under the umbrella of infectious interactions since they involve protein molecules of infectious bacteria with a pronounced specificity for certain cell surface components. Colicins are a well studied group of bacterial, plasmid coded proteins from *E. coli* [121,122]. They are hydrophilic molecules of 10^4 to 10^5 daltons and are especially interesting since as few as 1 to 10 molecules can have dramatic effects on a susceptible bacterium. Experiments using fluorescence probes reveal that binding of a few molecules to colicin can induce major membrane conformational changes or transitions. This observation provides a useful parallel with the effects of growth hormone on erythrocytes where binding of about 100 molecules produces similar changes throughout the entire cell surface [123].

Several types of colicins exist with diverse intracellular effects, including interference with protein synthesis, arresting replication and blocking active transport. The receptors for colicins, like phage receptors, are components of both cell wall and outer membrane. Their detailed chemistry has however not yet been elucidated. One fascinating aspect of colicin biology is the existence of both resistant and immune bacteria. The former probably lack colicin receptors whereas the latter apparently produce a reversible colicin antagonist which is coded for by the same plasmid as the colicin itself.

Bacterial toxins are of course of considerable clinical importance and are in fact responsible for inducing characteristic disease manifestations (e.g. tetanus, cholera, diphtheria — see for example [124]). The basic bifunctional nature of these proteins was predicted by Ehrlich at the beginning of the century (see Chapter 2); they each are composed of polypeptide subunits which together constitute a binding component and a biologically active 'toxophore'. Cholera and tetanus toxins exert their clinical effects on intestinal cells and neuro-muscular junctions; however, the receptors turn out to be fairly common membrane constituents — gangliosides

[125,126] (charged glycolipids). Cholera toxin binds selectively to G_{M1} ganglioside (with specificity for the Gal-GalNAc terminal polar group) on almost all cell types tested including red cells, adipocytes, liver parenchymal cells, neurones and lymphocytes. It has aroused considerable interest recently since its principle effect in both clinical and experimental situations is to activate the plasma membrane enzyme adenyl cyclase and thus elevate cyclic AMP levels [127]. The manner in which this is achieved is uncertain. It has been suggested that lateral mobilization of toxin-G_{M1} complexes within the phase of the membrane may facilitate a regulatory interaction with adenyl cyclase [128,39]. Alternatively, the mode of action may be less direct and involve internalization of the toxin and splitting off of the toxic moiety — perhaps parallelling what seems to happen with diphtheria toxin and the plant toxin, ricin [129]. Whatever its precise molecular *modus operandi*, cholera toxin promises to provide an intriguing system for analysing membrane activation phenomena, particularly since exogenous purified G_{M1} can be artificially inserted into cells (e.g. human leukaemia cells, virus transformed animal cells) deficient in this structure and therefore resistant to the effects of cholera toxin [39,40].

7.3 Developmental cell interactions

Understanding the molecular biology of embryogenesis provides the most formidable challenge to biologists. A complex myriad of cell interactions and associations are involved [130] and as Paul Weiss said in a review 20 years ago [131] — 'sheer compilation (of cell interactions) would produce more bewilderment than enlightenment'. When this diversity of interactions is considered alongside the dynamics of time and space involved, then it would seem that any reasonably complete comprehension is likely to elude us for some

time to come. In Chapter 2, I outlined the historical development of ideas on the specificity of embryological cell interactions and we saw that several outstanding 'heavyweights' of biology have emphasized the likely significance of stereospecific recognition by receptor molecules. Two other books in this series have dealt with broader problems of development and differentiation [132,133], and in this section, I shall briefly outline a general philosophy of cell surface phenomena in embryogenesis, review two analytical systems that promise to eventually provide us with detailed molecular biology of cell associations, and present some of the current and necessarily incomplete semimolecular models of cell interaction that have recently been put forward.

7.3.1 Selectivity and molecular specificity

Probably every decisive step in embryogenesis involves cell surface interactions and many of these certainly appear to be both selective and invariant or stereotyped [130,131]. These multiple interactions are however likely to involve different molecular mechanisms with varying degrees of specificity. There may be some value therefore in attempting to analyse cell surface phenomena in the context of distinctly different components of embryogenesis:

(1) Morphogenetic movements (polarity) and cell migrations (particularly germ cells and the presumptive pigment cells and nerve cells of the neural crest).

(2) Specification or differentiation of cells in terms of sequential genetic restriction of potentialities.

(3) Induction phenomena (e.g. neural plate, lens. Mesenchynal — epithelial interactions) which in most cases appear to be elicitive influences on pre-committed cells rather than instructive effects (cf. effects of antigens on lymphocytes).

(4) Histogenesis — isologous association or self-recognition of identical cells (cf. vascular anastomoses) and organogenesis — the association of dissimilar cells (e.g. adrenal cortex and medulla).

(5) Functional coupling (e.g. nerve end organ associations, retino-tectal connections).

(6) Functional regulation (e.g. by hormones). 3, 5, and 6 may differ in principle from 4 since they would seem to involve non-self or heterotypic interaction. One intuitively feels that there are therefore better candidates for *complementary receptor* based interactions although this need not necessarily be the case.

Whilst it is certain in some systems and likely in others that functional coupling and regulation involve highly discriminatory cell surface receptors, it is perhaps rather unlikely that it is true for the early crucial events in embryogenesis, particularly since *selective* expression of appropriate receptors would in essence beg the question of how these cells became specialized in the first place. For this reason, the idea that fundamental developmental restrictions are imposed on cells in accord with their field position (which may be largely fortuitous) and in response to *common* environmental cues, is a far more simple and attractive concept [134]. If this were to be correct then a small number of common receptors would be required and the differential programming induced would reflect quantitative and temporal aspects of the stimuli rather than qualitative signal discrimination. Recent studies on the role of cyclic AMP in the life of the slime mold *Dictyostelium* have suggested a possible molecular basis for such an arrangement (see Chapter 4). In these and other aspects of embryogenesis it is important to appreciate that apparent selectivity of stereotyped behaviour or cellular association need not necessarily reflect the existence of highly specific surface receptors. Cell contact like associations of amphipathic molecules in

solution is probably governed to a large extent by basic thermodynamic rules (maintenance of lowest free energy condition). Cells can clearly adhere perfectly well to non-biological surfaces and moreover in doing so they use much the same forces of attraction as employed in cell-cell adhesion (particularly electrostatic [135]).

When we consider this information along with the restrictive effects of time and space in embryogenesis then it seems perfectly reasonable to argue — as Curtis in particular has done [136] — that many selective cell interactions in embryogenesis may occur independently of distinct cell surface recognition events. This theoretical argument is supported by Weston's work on the neural crest which has strongly suggested that essentially non-specific environmental cues plus physical barriers may be sufficient to account for the remarkable migration pattern and selective localization of neural crest cells [137]. Such a non-specific element need not necessarily prejudice accuracy of the final outcome, neither does it deny that the activity of certain specialized cell surface components are involved. Cells *might* associate via interaction of common surface components or via specialized adherence sites. In the latter case which might for example be pertinent to the autologous (self) affinity tissue cells (see p. 62) the molecular mechanism involved might involve a highly discriminatory and perhaps sterically complementary interaction. However, there is no *a priori* reason to suppose that the structures involved are true *receptors* as opposed to specialised 'grappling hooks'. The same is true for the highly discriminatory interactions of microbial gametes mediated by mating type factors [104].

Some cellular interactions in embryogenesis probably involve cell surface structures that are true receptors in the sense that they both interact with or bind in a regulatory ligand (cell bound or free) *and* initiate a cellular response,

and clearest evidence for this so far comes from recent studies using a mesenchymal factor (a glycoprotein) which was covalently bound to agarose beads [138]. Pancreatic epithelia cells bind to these liganded beads (on their 'naturally' interacting basal sides) and were induced to both divide and develop into either exocrine acinar cells or endocrine B cells. A similar inductive phenomena is also seen with insolublilized Nerve Growth Factor [139]. It is important to appreciate, however, that these are essentially elicitive effects on cells that are already differentiated and in this sense the responses are no different in principle or more (or less) spectacular than for example the functional activation of resting lymphocytes by insolubilized ligands or responses of target cells to growth hormone or steroid sex hormones.

We might reasonably anticipate therefore that eventually embryological cell surface inter-actions will be shown to encompass a wide spectrum of molecular mechanisms ranging from the non-specific to highly discriminatory true receptor action.

The adoption of one or more of these various 'selective' devices by any particular interacting system may be determined primarily by physical factors of time and space. This is the essence of the problem in so far as specif-icity may only be *required* where multiple choice of association exists (in time and/or space) and random coupling is likely to be undesirable or inefficient. The clearest example of *specific* cell association would appear to be stereotyped synaptic connections in nerve tissue. Yet, the potential functional contact repertoire or real discriminatory capacity of individual neurons is a complete unknown. We at least suspect from studies involving surgical manipulation and transplantation that they are more flexible in their connections than their normal behaviour appears to suggest and that a one gene one synapse degree of specification is highly improbable (see p. 64).

7.3.2 Selective histiotypic associations

Holtfreter and later Moscona and colleagues have pioneered the study of the reaggregation in vitro of cells from vertebrate (particularly chick) embryo tissues [140,141,142]. This approach parallels earlier and current work on sponge cell reaggregation – and observations of tissue regeneration *in vivo*. All three systems involve preferential association of similar or identical cells (histiotypic or homotypic affinity). More dramatic re-associations have also been demonstrated *in vitro* involving a higher level of cell association and sorting. For example, heterogeneous dissociated cell populations from late embryonic cerebrum will not only reaggregate *in vitro* but also orientate themselves to establish a radially symmetrical tissue pattern resembling immature cerebral cortex. The basic assumption of such studies is that they may tell us something about *morpho-genesis* and may in particular lead to an understanding of the molecular basis of selective cell associations.

The sponge cell reaggregation system was introduced by Wilson at the beginning of the century, and studied extensively in the Twenties, by Galtsoff. Considerable further research over the past decade has been carried out by Humphreys. Maclennan, Burger and others. The older concept of species specific sorting of cells has given way to a more reasonable concept of selective sorting inter-actions at various taxonomic levels (from subspecies to orders [143]. However, the idea that preferential sorting in either sponge or vertebrate embryo systems must *necessarily* involve selective affinities has been chal-lenged [136]. The cells used in these assays have been dissociated by means (enzymatic, mechanical, calcium ion chelation) which inevitably distort the surface, and the analysis is often performed either on sick (abnormal) cells or cells which are in the process of regenerating cell surface components. Furthermore, there is some evidence to suggest that in sponge reaggregation at least an element of *mutual antagonism* may play a crucial role in effecting the apparent specificity of association. The suggestion should therefore be considered that some *homologous* interactions of cells could be based on non-specific *permissive* interactions whose apparent selectivity reflects time and/or space factors *or* discrimination against non-self.

The strongest argument for specific complementary cell surface interactions in reaggregation systems comes from the analysis of macromolecular cell products of both sponges [144,145] and vertebrate embryo cells [146,147]. These products which are now moderately well characterized (Table 7.2) have one crucial and intriguing quality -- they induce tissue specific (e.g. chick retina or cerebrum) or taxonomic (e.g. sponge species) selective reaggregation. We assume, therefore, that these macromolecules are functionally polyvalent cell surface reactive ligands. Moscona has proposed the general thesis that these *tissue specific* cell surface ligands are responsible for cell—cell recognition [140]. There is an obvious parallel here with the studies on mating type cell surface factors in micro-organisms (see p. 52), with at least one important difference — the latter are clearly a reflection of some form of molecular complementarity between non-homologous cells, i.e. it is a difference (or non-self) which is being recognised. There are also 'immunological' overtones in these studies since the occurrance of homotypic selectivity parallels the existence of tissue specific differentiation antigens [148]. Could these immunologically defined antigenic sites be *responsible* for selective associations? The

Table 7.2

Hausman and Moscona [146]	Balsamo and Lilien [147]
1. *Origin* Embryonic neural retina	Embryonic neural retina and cerebral lobe
2. *Assay* Tissue specific reaggregation	Tissue specific reaggregation
3. *Factor* (i) Glycoprotein 50 000 daltons (10—15% carbohydrate) (ii) Function carbohydrate *independent* (NB. No galactosyl transferase or sugar acceptor activity)	Glycoprotein Function carbohydrate *dependent* Retina: terminal N-acetyl-galactosamine Cerebral cells: terminal mannoside (binding inhibited by glycosidases or addition of free sugars)
4. *Co-factors required for function*	3rd 'ligator' component (see Fig. 7.) (cf. Aggregation Factor in Burger model — Fig. 7.)

Note: Aggregation promoting factors from sponges appear to be polydisperse glycoproteins whose activity is sensitive to glycosidases [144, 145].

antigenically identifiable mating type factors in micro-organisms and phage receptors on bacteria provide other possible precedents. Supporting evidence comes from immunological studies on the H-2 system of mice. This locus includes genes coding for cell surface determinants involved in interactions in the immune response (see Chapter 6). It is also closely linked to the T locus which exerts profound developmental effects involving immunological detectable cell surface structures [148].

There are two interesting variants of the reaggregation factor experiments. Glaser and colleagues [149] have prepared radioactive chick embryo retina and cerebella plasma membrane fragments which bind selectively (in a simple two way reciprocal test) to their cells of origin and *block* reaggregation. These experiments also revealed suggestive temporal as well as tissue specificity since membranes prepared from retina at different ages (e.g. seven through 12 days) of embryogenesis appeared to preferentially inhibit reaggregation of cells of the same age as the membrane 'donor'. Burger and colleagues have developed a cell-free assay system for sponge aggregation factors [144]. Proteins from hypotonically shocked cells have been covalently bound to agarose beads. The treated beads bind to mechanically disrupted cells and are themselves agglutinated by sponge cell supernatant aggregation factors. The specificity of these reactions and the chemistry of the factors involved obviously require a great deal more study. They do, however, along with other variants of cell binding assays [150], provide a more realistic quantitative means of investigating discriminatory cell surface reactions.

7.3.3 Retino-tectal specificity

The nervous system in terms of cellular heterogeneity and the complexity of its integrated function provides on the one hand the most difficult system to analyse and on the other the system which is perhaps most likely to reveal cellular interactions based on complementary receptors. Homotypic associations as studied by Moscona and colleagues in nervous tissue may be an important ingredient of development; however, the selective association and functional coupling between *different* cells is in many respects more intriguing since it represents recognition of non-self and appears intuitively as a more promising candidate for specific receptor function.

The invariance of certain synaptic connections and pattern formation and their re-establishment after surgical manipulation or disaggregation *in vitro* has suggested that some form of *neuronal coding or specificity* exists — probably expressed in cell surface structures [151]. Two systems in particular have been extensively studied and discussed — nerve end organ interaction and specification and spatial aspects of retina-optic tectum association. Both of these approaches were pioneered by Sperry in the Forties [152]. His general approach was to disrupt 'normal' connections and to observe the selectivity of re-innervation. Sperry's early studies on skin sensitization involved transplanting skin from one part of a frog to another (e.g. belly to back) and he interpreted the 'reversed scratching responses' observed to indicate that the same nerve which was severed regrew back to its old partner. More recent studies (e.g. by Miner, Szekely and Gaze [153,154]) have however shown that there is probably *no* selective outgrowth to the 'correct' regions. Sperry's most important contributions to neurological specificity were his observations on retino-tectal specificity [153]. In his original experiments he severed the optic nerve of a frog tadpole and rotated the eye 180°. Re-innervation occurred and the adult frog showed *inverted* visual responses. Sperry concluded that the severed optic nerves made their way back to the correct position in the optic tectum, and that this space coded specificity was controlled by some 'chemical' specificity of the cells involved. A very considerable amount of effort has since been expanded in attempts to verify this hypothesis and to date, no definitive proof or disproof exists. What seems to have emerged from studies in amphibians and chickens using developmental, regenerative, anatomical, and electrophysiological approaches is a rather more flexible picture. Prior to stage 30 (in chicks) very little precise specification appears to exist. Post stage 30, specific connections are indeed established. However, they are probably preceded by considerable make and break or trial and error interaction before a final stable interconnection is established. Moreover, surgical studies imply that a considerable capacity for reorganization or regulation may exist even at late stages. On the basis of these various experiments, Gaze, Keating and colleagues have re-interpreted Sperry's and their own earlier findings in terms of a flexible 'systems matching' model [154]. Not surprisingly, this topic is surrounded by a healthy air of controversy and the interested reader would benefit from reading lucid reviews on this subject by Gaze and Keating [154], Chung and Feldman [156] and Hunt and Jacobson [153]. The latter authors in particular have critically evaluated the *assumption* that the test systems used are in fact intrinsically capable of revealing precise specificity should it exist. They point out several clear deficiencies and make several important suggestions for experiments to test the real *discriminatory* power of optic neurons by, for example, *competitive* innervation. Recent work (reviewed in [157]) strongly suggests that competitive innervation is demonstrable and in all likelihood this may provide us with the best evidence for specificity in synapse formation. In future studies it will certainly be important to distinguish what may be transient 'probing' contacts from definitive

functional coupling. It is difficult to imagine that this will ever be achieved *in vivo*. Fortunately, semi-quantitative systems are now being developed for analysing retino-tectal cell interactions *in vitro*. Roth and colleagues have reported a selective binding of radio-labelled cells from either dorsal or ventral retina to the cells from the corresponding reciprocal half of the optic tectum [142]. The results, although statistically significant, were not overly impressive in terms of selective binding. Nevertheless, this type of system should provide the opportunity to move towards an understanding of the molecular basis of neurological specificity.

An alternative approach to selective interactions and synaptogenesis in the nervous system has been introduced by Sidman and colleagues from Harvard Medical School [158]. The approach is to search for *gene* controlled events which regulate cell behaviour. A similar gene-function approach is also being applied to other embryonic systems (e.g. neural crest [137]). Two mice mutants have been found which show abnormalities in the brain. In one case ('staggerer' mice), there appears to be a gross deficiency of synaptic association between granule cell axons and Purkinje dendrites. In another ('reeler' mice), there is a marked and stereotyped abnormality in topographical pattern of cerebellar and cerebral organization. Reaggregation experiments show that the abnormality, whatever its basis, is also reflected in the way 'reeler' brain cells interact *in vitro*. Sidman has suggested that this disorder arises early in embryonic life as a single gene effect and may be expressed by cell surface structure(s) involved in cell associations and pattern formation.

7.3.4 Molecular models
In the light of our current state of knowledge detailed molecular models for selective cell association in embryogenesis are clearly

premature, however, general models may have a considerable heuristic value in suggesting future experimental approaches. There are two fundamental questions which need to be posed in constructing such models:

(1) what is the overall symmetry of the recognition phenomenon, and

(2) what is the likely general type of specificity involved.

With respect to the first question there would seem to be three general arrangements possible as shown in Fig. 7.4. I have deliberately placed Weiss' concepts published in 1947 alongside those of current exponents of the art. Burger's model is derived from experimental studies on sponge cell reaggregation [144] and that of Balsamo and Lilien [159] from studies on neural retina cell reaggregation.

The nature of the specificities involved is of course a completely open question. One particular avenue which seems worth exploring is that cell surfaces may have lectin-like proteins which can selectively bind different oligosaccharides (on other cell surfaces). True lectins have been isolated primarily from plants of the Leguminosae family but strikingly similar anti-saccharide proteins exist in invertebrates (e.g. sponge extracts, snail haemolymph). There is preliminary evidence that the sponge reaggregation factors are glycoproteins and that their activities are inhibited or destroyed by interaction with glycosidases and some lectins [144]. Burger also quotes preliminary studies which suggested that oligosaccharides derived from sponge (*Microciona prolifera*) aggregation factor function as 'haptens' in inhibiting reaggregation [144]. Roth and colleagues have also reported that carbohydrates perturb neural retina cell reaggregation — evidence which they take to support a different kind of model.

An alternative to the 'lectin' model, but one which is extremely similar in principle has been proposed by Roseman and developed by Roth in particular (reviewed in [142]). The principle involved in this model is one of complementary

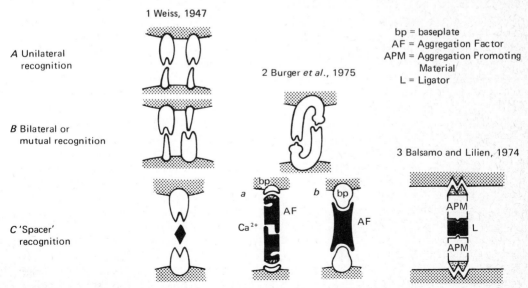

Fig 7.4 Some molecular models for cell—cell interaction

affinity with an *enzyme* as the receptor, as shown in Fig. 7.5. It is suggested that glycosyl transferases on cell surfaces interact with other cells via sugar acceptor groups. This imaginative and important hypothesis has been critically reviewed by Roth and I shall only mention here a few important points.

Glycosyl transferases are indeed found on the surface of four cell types studied: rat intestinal epithelial cells, mouse embryo fibroblast cell lines, chick neural retina cells and human blood platelets. A recent report suggests that glycosyl transferase activity is greatly increased on flagellae of *Chlamydomonas* when compatible mating pairs are mixed together, implying that this enzyme is directly or indirectly involved in the coupling reaction [160]. Glycosyl transferases have been reported to be absent from lymphocytes which may apparently have fucosyl transferases!

Direct evidence that these enzymes are involved in or responsible for cellular adhesion is at present lacking, although this seems likely to be the case at least in the binding of platelets to collagen. In the latter instance, it has been shown that platelets transfer glucose to collagen glycopeptides with exposed galactose residues and this activity is inhibited by agents which suppress platelet-collagen adhesion (e.g. aspirin, chlorpromazine and glucosamine). Mouse fibroblasts *in vitro* are also capable of catalysing the transfer of galactose from UDP-Gal to either cellular or soluble acceptors.

The similar enzyme-sugar recognition system in the adult rat liver has been described by Ashwell and colleagues [161]. They have shown that removal of circulating glycoproteins by liver cells is strongly dependent upon their terminal sugar residues. If native glycoproteins have their terminal sialic acid residues removed and galactose residues exposed, then they too are rapidly removed by the liver or bound by liver cell membranes *in vitro*. Significantly, if the liver membranes, which presumably have a

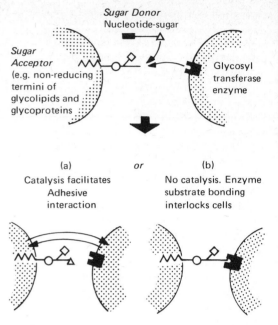

Fig. 7.5 The glycosyl transferase model for cell interaction

galactose recognition site, are treated with neuraminidase they can no longer bind asialoproteins. Binding activity can be restored by incubating membranes with cytidine-monophosphate sialic acid, which is a sialic acid donor. One interpretation of these intriguing studies is that the galactose receptor is an sialotransferase enzyme, whose activity against circulating or free asialoproteins can be inhibited by sialidase induced production of endogenous asialcompounds. The idea that enzyme-substrate interactions are involved in cellular affinities is an important one which will no doubt be extensively investigated in the next few years. It is perhaps worth recalling here that enzymes had a brief and unsuccessful run as candidate receptors for drugs and neurotransmitters. A phylogenetic or evolutionary link between enzymes and receptors nevertheless deserves serious consideration particularly since many enzymes function as membrane bound molecules and incorporate the two key qualities one would like to see in a receptor — namely specificity and regulatory activity.

Epilogue

Throughout this review of receptor biology, I have stressed both the diversity of intercellular recognition systems and their fundamental importance for biological processes and control. Inevitably in a book of this size many topics have been dealt with superficially and others left out entirely. I should particularly mention *direct* intercellular communication and also chemoreceptors. The former is manifested by electrical and metabolic coupling of cells via 'gap' or 'tight' junctions and may be an important role both in development and normal physiological responses [162,163]. Chemoreception from bacteria to man is a fascinating area and considerable advances have been made in isolating receptors from microorganisms [164]; the great diversity of environmental smells and specific pheromones (and tastes) that can be distinguished at very low concentrations [6] implies the existence of a corresponding variety of highly discriminatory receptors — an interesting parallel here with the immune system [5]. The molecular basis of odour discrimination is unknown; stereospecific recognition has been postulated [165] and some direct evidence for this exists [166].

I have also tried to emphasize the contrast between the enormous variety of chemical signals interacting with corresponding cell surface receptors and the existence of an apparently common effect of ligand-receptor binding on membrane properties and cellular response. This principle of parsimony is bound to be an oversimplification but it has a logical simplicity from the evolutionary standpoint. Specificity determined by structural genes is essential both for characteristic cellular response, whatever it may be (e.g. contraction, hormone secretion), and for receptors which provide *access* for extracellular signals which are essential for regulating specialized activity. In contrast, there is really no need for each receptor-response pathway to involve individually specific coupling devices. This possibility rests heavily on the premise that almost all intracellular signals which regulate cell behaviour are elicitive rather than specifically instructive. They simply tell a cell to do more or less what it is intrinsically programmed to do. The signals from the receptor to the biosynthetic and metabolic machinery of the cell can therefore afford to be relatively common or pleiotypic. It is reassuring in this context to find that both basic plasma membrane structure and behaviour and common coupling mechanisms, e.g. cyclic nucleotides, calcium ions, are found in microbial organisms and throughout the animal world. This reminds one of Francois Jacob's comparison of *E. coli* and elephants. Clearly once the genetic code had been 'invented' its basic blueprint was retained. Cell surface recognition events, although more diverse, may also have had sufficient selective advantage to be maintained throughout evolution in at least a recognizably common form.

There are also strong hints that receptors themselves may have much more in common than one might suspect from their diverse locations and responses they elicit. Stereospecific binding by cell surface proteins is clearly the central characteristic or paradigm. Binding specificity for saccharides is also a re-occurring theme in widely distinct cellular interacting systems. Cell surface glycoproteins, identified immunologically as products of genes within or linked to the major histocompatibility locus of the species, may well play a crucial role in a great variety of recognition systems.

Analogies between fertilization, cellular

interactions in development and immunological reactions, although of historical value and interest, might now be regarded as too superficial. I suggest that when the cell surface localized structures and molecular mechanisms underlying these basic biological phenomena are unravelled, the analogy will however be strengthened and a surprising degree of homology revealed.

SUGGESTIONS FOR FURTHER READING

Recommended books and reviews on recognition and receptors.

The Cell Periphery, Metastasis and Other Contact Phenomena. (1967), Weiss, L. Academic Press, London.
Structure and Function of Biological Membranes. (1971), Rothfield, L. I. (ed.) Academic Press, London and New York.
Biological Membranes, Vol. 2. (1973), Chapman, D. and Wallach, D. F. H. Academic Press, London and New York.
Cell Interactions. (1972), Silvestri, L. G. (ed.) North-Holland, Amsterdam.
Neurochemistry of Cholinergic Receptors. (1974), De Robertis, E. and Schacht, J. (eds.) Raven Press, New York.

Cell Surface Receptors. (1975), Nicolson, G., Raftery, M. and Rodbell, M. (eds.) UCLA-ICN Winter Conference, Academic Press, London and New York.
Receptors and Recognition. (1975), Cuatrecasas, P. and Greaves, M. F. (eds.) Chapman and Hall, London.

Prominent journals, publishing reviews and new research papers on biological recognition and receptors.

Annual Reviews of Biochemistry, Microbiology and Pharmacology; Progr. in Biophys.; Adv. Cyclic Nucleotide Res.; Nature; Science; Proc. Nat. Acad. Sci. (US); J. Membrane Biol.; Cell; Exp. Cell Res.; Biophys. Biochem. Acta; J. Biol. Chem.; Biochem. Biophys. Res. Commun.; J. Molec. Biol.; J. Exp. Med.

References

[1] Bernard, C. (1957), *An Introduction to the Study of Experimental Medicine,* Dover, New York.
[2] Michener, C.D. and Brothers, D.J. (1974), *Proc. Nat. Acad. Sci.* (US), **71**, 671.
[3] Weiner, N. (1961), *Cybernetics or Control and Communication in the Animal and Machine,* Wiley, New York.
[4] *Neuroscience Research Programme Bulletin,* (1970), **8**, 4.
[5] Thomas, L. (1974), *Progr. Immunol. II,* North-Holland, Amsterdam, p. 239.
[6] Birch, M.C. (ed.) (1974), *Pheromones,* North-Holland, Amsterdam.
[7] King, R.J.B. and Mainwaring, W.I.P. (1974), *Steroid-Cell Interactions,* Butterworth, London.
[8] Perkins, J.P. (1973), *Adv. Cyclic Nucleotide Res.,* 3, 1.
[9] Von Uexkull, quoted in Tinbergen, N. (1951), *The Study of Instinct,* Clarendon Press, Oxford.
[10] Rodbell, M. (1972), In *Current Topics in Biochemistry,* Academic Press, New York, p. 187.

[11] Paston, I. and Perlman, R.L. (1971), *Nature New Biol.,* **229**, 5.
[12] Goldberg, N.D., O'Dea, R.F. and Haddox, M.K. (1973), *Adv. Cyclic Nucleotide Res.,* **3**, 155.
[13] Tamar, H. (1972), *Principles of Sensory Physiology,* C.C. Thomas, Springfield, Illinois.
[14] Miller, W.H., Gorman, R.E. and Bitensky, M.W. (1971), *Science,* **174**, 295.
[15] Jacob, F. (1974), *The Logic of Living Systems,* Allen Lane, London.
[16] Pauling, L. (1960), *Nature of the Chemical Bond,* Cornell Univ. Press, Ithaca.
[17] Koshland, D.E. and Neet, K.E. (1968), *Ann. Rev. Biochem.,* **37**, 359.
[18] Burgen, A.S.V., Roberts, G.C.K. and Feeney, J. (1975), *Nature,* **253**, 755.
[19] Changeux, J.P., Thiery, J., Tung, Y. and Kittle, C. (1967), *Proc. Nat. Acad. Sci.* (US), **57**, 335.
[20] *The Collected Papers of P. Ehrlich,* (1957), Vol. 2, Pergamon Press, London, p. 178.
[21] Burnet, F.M. (1959), *The Clonal Selection Theory of Acquired Immunity,* Cambridge Univ. Press, London.

[22] Landsteiner, K. (1936), *The Specificity of Serological Reactions,* Dover, New York.
[23] Haurowitz, F. (1962), *Biol. Rev., 27,* 247.
[24] Rang, H.P. (1971), *Nature, 231,* 91.
[25] Sela, M. (1972), *Harvey Lectures, 67,* 213.
[26] Zabriskie, J. (1967), *Adv. Immunol., 7,* 147.
[27] Tyler, A. (1946), *Growth,* 10, 7.
[28] Weiss, P. (1947), *Yale Jol. Biol. Med., 19,* 235.
[29] Monod, J. (1972), *Chance and Necessity,* Collins, London.
[30] Pauling, L. (1956), In *Enzymes: Units of Biological Structure and Function,* Academic Press, New York, p. 177.
[31] Pauling, L. (1974), *Nature, 248,* 769.
[32] Wallach, D.F.H. (1972), *The Plasma Membrane* Engl. Univ. Press, London.
[33] Singer, J. (1974), *Ann. Rev. Biochem., 43,* 805.
[34] Singer, J. and Nicolson, G. (1972), *Science,* 175, 720.
[35] Marchesi, V.T., Tillack, T.W., Jackson, R.L., Segrest, J.P. and Scott, R.E. (1972), *Proc. Nat. Acad. Sci.* (US), 69, 1445.
[36] Lis, H. and Sharon, N. (1973), *Ann. Rev. Biochem., 43,* 541.
[37] Nicolson, G. (1974), *Int. Rev. Cytol., 39,* 89.
[38] Edidin, M. (1974), *Ann. Rev. Biophys. Bioeng.,* 3, 179.
[39] Frye, L.O. and Ediden, M. (1970), *J. Cell Sci.,* 7, 319.
[40] Raff, M.C. and de Petris, S. (1973), *Fed. Proc.,* 32, 1.
[41] Sutherland, E.W. and Robinson, C.A. (1966), *Pharm. Rev., 13,* 145.
[42] Gerische, G. and Hess, B. (1974), *Proc. Nat. Acad. Sci.* (US), 71, 2118.
[43] Epstein, P. (1974), *Nature, 251,* 572.
[44] Birnbaumer, L. (1973), *Biochim. Biophys. Acta,* 300, 129.
[45] Cuatrecasas, P. (1974), *Ann. Rev. Biochem., 43,* 169.
[46] Ramwell, P. (ed) (1974), *Prostaglandins,* Plenum Press, London.
[47] Goldberg, N.D., Haddox, M.K., Dunham, E. *et al.* (1974), In *Control of Proliferation in Animal Cells,* Cold Spring Harbor Press, New York., p. 609.
[48] Seifert, W.E. and Rudland, P.S. (1974), *Nature,* 248, 138.
[49] Rasmussen, H., Goodman, D.B.P. and Tanenhouse, A. (1972), *CRC, Critical Rev. Biochem.,* 1, 95.
[50] Rubin, R.P. (1970), *Pharmacol. Rev., 22,* 389.
[51] Cone, R.A. (1975), In *Perspectives in Membrane Biology,* Academic Press, New York.
[52] Maino, V., Green N.M. and Crumpton, M.J. (1974), *Nature, 251,* 324.
[53] Foreman, J.C., Mongar, J.L. and Gomperts, B.D. (1973), *Nature, 254,* 249.
[54] Selinger, Z., Eimerl, S. and Schramm, M. (1974), *Proc. Nat. Acad. Sci.* (US), 71, 128.
[55] Steinhardt, R.A., Epel, D., Carroll, E.J. and Yanagimachi, R. (1974), *Nature, 252,* 41.
[56] Freedman, M.H., Raff, M.C. and Gomperts, B.D. (1975), *Nature, 255,* 378.
[57] Greaves, M.F. and Janossy, G. (1972), *Transpl. Rev., 11,* 87.
[58] Resch, K. (1976), In *Receptors and Recognition* Vol. 1, Chapman and Hall, London.
[59] Albert, A. (1971), *Ann. Rev. Pharmacol., 11,* 13.
[60] Katz, B. (1966), *Nerve, Muscle and Synapse,* McGraw Hill, USA.
[61] Daniels, M.P. and Vogel, Z. (1975), *Nature,* 254, 339.
[62] Cohen, J.B. and Changeux, J.-P. (1975), *Ann. Rev. Pharmacol., 15,* 83.
[63] Rang, H.P. (ed) (1973), *Drug Receptors,* Macmillan, New York.
[64] Axelrod, J. (1974), *Neurotransmitters,* Scientific American, p. 59.
[65] Michelson, M.J. (1974), *Biochem. Pharmacol.,* 23, 2211.
[66] Michaelson, D.M. (1975), In *Cell Surface Receptors,* ICN-UCLA Winter Conference, Academic Press, New York.
[67] Weight, F.F., Petzold, G. and Greengard, P. (1974), *Science,* 186, 942.
[68] Karlin, A. (1973), *Fed. Proc., 32,* 1847.
[69] Triggle, D.J. (1971), *Neurotransmitter-Receptor Interactions,* Academic Press, New York.
[70] Atlas, D., Steer, M.L. and Levitski A. (1974), *Proc. Nat Acad. Sci.* (US), 71, 4246.
[71] Triggle, D.J. (1972), *Ann. Rev. Pharmacol., 12,* 185.
[72] Singer, S.J., Ruoho, A., Kiefer, H., Lindstrom, J. and Lennox, E.S. (1973), In *Drug Receptors,* Academic Press, New York, p. 183.
[73] Roberts, E. and Matthysse, S. (1970), Academic Press, New York.
[74] *Receptor Biophysics and Biochemistry,* Neurosciences Res. Program Bulletin, *11* (3), M.I.T. Press, USA (1973).
[75] Cuatrecasas, P. (1974), *Biochem. Pharmacol.,* 23, 2353.
[76] Venter, J.C., Dixon, J.E., Maroko, P.R. and Kaplan, N.O. (1972), *Proc. Nat. Acad. Sci.* (US), 69, 1141.

[77] Greaves, M.F. and Bauminger, S. (1972), *Nature New Biol.*, **235**, 67.

[78] Kolb, H.J., Renner, R., Hepp, K.D., Weiss, L. and Wieland, O.H. (1975), *Proc. Nat. Acad. Sci.* (US), **72**, 248.

[79] Wilchek, M., Oka, T. and Topper, Y.J. (1975), *Proc. Nat. Acad. Sci.* (US), **72**, 1055.

[80] Blundell, T.L., Hodgkin, D.C., Dodson, E., Dodson, G.G. and Vijayan, M. (1971), *Rec. Prog. Hormone Res.,* **27**, 1.

[81] Greaves, M.F., Owen, J.J.T. and Raff, M.C. (1973), *T and B lymphocytes,* Excerpta Medica, Amsterdam.

[82] Porter, R. (ed) *Defence and Recognition,* M.T.P. Int. Rev. Sci. (Biochem.), Vol. 10.

[83] Moller, G. (ed) (1973), *Transpl. Rev,* Vol. 14.

[84] Cooper, E.L., (ed) (1974), *Contemporary Topics in Immunobiology,* Vol. 4, Plenum Press, New York.

[85] Klein, J. and Shreffler, D.C. (1971), *Transpl. Rev.,* **6**, 3.

[86] Bodmer, W.F. (1972), *Nature,* **237**, 139.

[87] Sachs, D.H. and Dickler, H.B. (1975), *Transpl. Rev.,* **25**, 159.

[88] Benacerraf, B. and McDevitt, H.O. (1972), *Science,* **175**, 273.

[89] Burnet, F.M. (1971), *Nature,* **226**, 123.

[90] Hildemann, W.H. and Reddy, A.L. (1973), *Fed. Proc.,* **32**, 2189.

[91] Moller, G. (ed) (1974), *Transpl. Rev.* Vol. 21.

[92] Greaves, M.F. (1975), In *Immune Recognition,* Rosenthal, A. (ed) Academic Press, New York, p. 3.

[93] Andersson, J., Sjöberg, O. and Möller, G. (1972), *Transpl. Rev.,* **11**, 131.

[94] Möller, E. (ed) (1975), *Transpl. Rev.,* Vol. 23.

[95] Feldmann, M., Greaves, M.F., Parker, D.C. and Rittenberg, M. (1974), *Eur. J. Immunol.,* **4**, 591.

[96] Becker, K.E., Ishizaka, T., Metzgar, H., Ishizaka, K. and Grimley, P.M. (1973), *J. Exp. Med.,* **138**, 394.

[97] Hadden, J.W., Johnson, E.M., Hadden, E.M., Coffrey, R.G. and Johnson, L.D. (1975), In *Immune Recognition*, Academic Press, New York, p. 359.

[98] Metz, C.B. and Monroy, A. (eds) (1969), *Fertilisation*, Vols. 1 and 2, Academic Press, New York.

[99] Koltin, Y., Stamberg, J. and Lemke, P.A. (1972), *Bact. Rev.,* **36**, 156.

[100] Lewis, D. (1965), In *Genetics Today*, **3**, 657, Pergamon Press, Oxford.

[101] Linskens, H.F. (1969), In *Fertilisation,* Vol. 2, Academic Press, New York, p. 189.

[102] Burnet, F.M. (1971), *Nature*, **232**, 230.

[103] Crandall, M., Lawrence, L.M. and Saunders, R.M. (1974), *Proc. Nat. Acad. Sci.* (US), **71**, 26.

[104] Reissig, J.L. (1974), *Current Topics in Microbiol. and Immunol.,* **67**, 44.

[105] Weise, L. (1969), In *Fertilisation,* Vol. 2, Academic Press, New York.

[106] Sermonti, G. (1969), In *Fertilisation,* Vol. 2, Academic Press, New York, p. 47.

[107] Burnet, F.M. (1934), *Biol. Rev. Biol. Proc., Cambridge Phil. Soc.,* **9**, 332.

[108] Lindberg, A.A. (1973), *Ann. Rev. Microbiol.,* **27**, 205.

[109] Kohn, A. and Fuchs, P. (1973), *Adv. in Virus Res.,* **18**, 159.

[110] Dales, S. (1973), *Bact. Rev.,* **37**, 103.

[111] Dvorak, J.A., Miller, L.H., Whitehouse, W.C. and Shiroishi, T. (1974), *Science,* **187**, 748.

[112] Miller, L.H., Dvorak, J.A., Shiroishi, T. and Durocher, J.R. (1973), *J. Exp. Med.,* **138**, 1597.

[113] Hughes, R.C. (1973), *Prog. Biophys. Mol. Biol.,* **26**, 189.

[114] Haywood, A.M. (1974), *J. Mol. Biol.,* **83**, 427.

[115] Aoki, T. and Takahashi, T. (1972), *J. Exp. Med.,* **135**, 443.

[116] Notkins, A.L., Mergenhagen, S.E. and Howard, R.J. (1970), *Ann. Rev. Microbiol.,* **24**, 525.

[117] Valdimarsson, H., Agnarsdottir, G. and Lachmann, P.J. (1975), *Nature,* **255**, 554.

[118] Greaves, M.F., Brown, G. and Rickinson, A.B. (1975), *Clin. Immunol. Immunopath.,* **3**, 514.

[119] Klein, G. (1972), *Proc. Nat. Acad. Sci.* (US), **69**, 1056.

[120] Miller, D.A., Miller, O.J., Dev, V.G., Hashmi, S. Tratravahi, R., Medrano, L. and Green, H. (1974), *Cell,* **1**, 167.

[121] Nomura, M. (1967), *Ann. Rev. Microbiol.,* **21**, 257.

[122] Luria, S.E. (1973), in *Bacterial Membranes and Walls,* Decker, New York.

[123] Sonenberg, M. (1971), *Proc. Nat. Acad. Sci.* (US), **68**, 1051.

[124] Finkelstein, R.A. (1973), *CRC Critical Reviews in Microbial.,* **2**, 553.

[125] Cuatrecasas, P. (1973), *Biochem.,* **12**, 3547.

[126] King, C.A. and Van Heyningen, W.E. (1973), *J. Infect. Dis.,* **127**, 639.

[127] Holmgren, J., Lindholm, L. and Lonnroth, I. (1974), *J. Exp. Med.,* **139**, 801.

[128] Bennet, V., O'Keefe and Cuatrecasas, P. (1975), *Proc. Nat. Acad. Sci.* (US), **72**, 33.

[129] Hughes, R.C. (1975), *Essays in Biochem.*, Vol. 2.

[130] Deuchar, E.M. (1975), *Cellular Interactions in Animal Development*, Chapman and Hall, London.

[131] Weiss, P. (1958), *Int. Rev. Cyt.*, 7, 391.

[132] Garrod, D.R. (1973), *Cellular Development*, Outline Studies in Biology, Chapman and Hall, London.

[133] Ashworth, J.M. (1973), *Cell Differentiation* Outline Studies in Biology, Chapman and Hall, London.

[134] Wolpert, L. (1969), *J. Theoret. Biol.*, 25, 1.

[135] Curtis, A.S.G. (1973), *Progr. Biophys. Mol. Biol.*, 27, 317.

[136] Curtis, A.S.G. (1972), In *Functional Aspects of Parasite Surfaces*, Blackwell Scientific Publications, Oxford.

[137] Weston, J.A. (1972), In *Cellular Interactions*, North-Holland, Amsterdam, p. 286.

[138] Levine, S., Pictet, R. and Rutter, W.J. (1973), *Nature*, 246, 49.

[139] Frazier, W.A., Boyd, L.F. and Bradshaw, R.A. (1973), *Proc. Nat. Acad. Sci.* (US), 70, 2931.

[140] Moscona, A.A. (1974), In *The Cell Surface in Development*, Wiley, New York, p. 67.

[141] Lilien, J.E. (1969), *Curr. Topics, Develop. Biol.*, 4, 169.

[142] Roth, S. (1973), *Quart. Rev. Biol.*, 48, 541.

[143] MacLennan, A.P. (1970), *Symp. Zool. Soc.* (London), 25, 299.

[144] Burger, M., Turner, R.S., Kuhns, W.J. and Weinbaum, G. (1975), *Phil. Trans. R. Soc.* (London) B, p. 147.

[145] Henkart, P., Humphreys, S. and Humphreys, T. (1973), *Biochemistry*, 12, 3045.

[146] Hausman, R.E. and Moscona, A.A. (1975), *Proc. Nat. Acad. Sci.* (US), 72, 916.

[147] Balsamo, J. and Lilien, J. (1975), *Biochem.*, 14, 167.

[148] Artzt, K. and Bennett, D. (1975), *Nature*, 256, 545.

[149] Merrill, R. and Glaser, L. (1973), *Proc. Nat. Acad. Sci.* (US), 70, 2794.

[150] Barbera, A.J., Marchese, R.B. and Roth, S. (1973), *Proc. Nat. Acad. Sci.* (US), 70, 2482.

[151] Gaze, R. (1970), *The Formation of Nerve Connections*, Acad. Press, New York.

[152] Sperry, R.W. (1945), *J. Neurophysiol.*, 8, 15.

[153] Hunt, R.K. and Jacobson, M. (1974), *Current Topics in Developmental Biol.*, 8, 203.

[154] Gaze, R.M. and Keating, M.J. (1972), *Nature*, 237, 375.

[155] Sperry, R.W. (1965), In *Organogenesis*, Holt, Rinehart and Winston, New York, p. 161.

[156] Chung, S.H. and Feldman, J.D. (1973), In *Biological Diagnosis of Brain Disorders*, Spectrum, Flushing, New York, p. 193.

[157] Horrdige, G.A. (1972), In *Cellular Interactions*, North-Holland, Amsterdam, p. 15.

[158] Sidman, R.L. (1972), In *Cellular Interactions*, North-Holland, Amsterdam, p. 1.

[159] Balsamo, J. and Lilien, J. (1974), *Nature*, 251, 522.

[160] McLean, R.J. and Bosmann, H.B. (1975), *Proc. Nat. Acad. Sci.* (US), 72, 310.

[161] Ashwell, G. and Morell, A.G. (1974), *Adv. Enzymol.*, 41, 99.

[162] Pitts, J.D. (1975), *Nature*, 255, 371.

[163] Azarnia, R., Larzen, W.J. and Loewenstein, W.R. (1974), *Proc. Nat. Acad. Sci.* (US), 71, 880.

[164] Adler, J. (1974), in *Chemotaxis its Biology and Biochemistry*, Antibiotics and Chemotherapy, Vol. 19, S. Karger, Basel.

[165] Amoore, J.E. (1970), *Molecular Basis of Odour*, C.C. Thomas, Springfield, USA.

[166] Getchell, M.L. and Gestland, R.C. (1972), *Proc. Nat. Acad. Sci.* (US), 69, 1494.

[167] Revesz, T. and Greaves, M.F. (1975), *Nature*, 257, 103.

[168] Cuatrecasas, P. (1973), *Biochem.*, 12, 3558.

[169] Poljak, R.J. (1975), *Nature*, 256, 373.

[170] Wahn, H.L., Lightbody, L.E. and Tchen, T.T. (1975), *Science*, 188, 366.

Index